Selected Papers from the 8th Symposium on Micro-Nano Science and Technology on Micromachines

Selected Papers from the 8th Symposium on Micro-Nano Science and Technology on Micromachines

Special Issue Editors

Norihisa Miki
Koji Miyazaki
Yuya Morimoto

MDPI • Basel • Beijing • Wuhan • Barcelona • Belgrade

MDPI

Special Issue Editors

Norihisa Miki
Keio University
Japan

Koji Miyazaki
Kyushu Institute of Technology
Japan

Yuya Morimoto
The University of Tokyo
Japan

Editorial Office
MDPI
St. Alban-Anlage 66
4052 Basel, Switzerland

This is a reprint of articles from the Special Issue published online in the open access journal *Micromachines* (ISSN 2072-666X) in 2018 (available at: https://www.mdpi.com/journal/micromachines/special_issues/MNST2017)

For citation purposes, cite each article independently as indicated on the article page online and as indicated below:

LastName, A.A.; LastName, B.B.; LastName, C.C. Article Title. *Journal Name* **Year**, *Article Number*, Page Range.

ISBN 978-3-03897-728-5 (Pbk)
ISBN 978-3-03897-729-2 (PDF)

Contents

About the Special Issue Editors

Norihisa Miki received his Ph.D. in mechano-informatics from University of Tokyo in 2001. He then worked at the MIT microengine project as a postdoctoral associate and later as a research engineer. He joined the Department of Mechanical Engineering of Keio University in 2004 as an associate professor and became a full professor in 2017. His research interests include micro/nano biomedical devices and information communication technologies (ICT). He was a JST PRESTO researcher from 2010 to 2016 and at the Kanagawa Institute of Industrial Science and Technology (formerly, Kanagawa Academy of Science and Technology). He was the general chair of the JSME 8th and 9th Symposia on Micro Nano Science and Technology in 2017 and 2018. He co-founded a healthcare startup, LTaste Inc., in 2017.

Koji Miyazaki received his Ph.D. in mechanical engineering and science from Tokyo Institute of Technology in 1999. He joined the Department of Mechanical and Control Engineering at Kyushu Institute of Technology in 1999 as a lecturer and became a full professor in 2011. His research interests include the thermophysical properties of nano-structured materials and micro thermal devices such as a thermoelectric micro-generators. He stayed at UCLA from 2000 to 2001 and at MIT from 2001 to 2002 as a visiting scholar. He was a JST PRESTO researcher from 2004 to 2008, and he is currently a PI of JST CREST, since 2017. He will be a general chair of the JSME 10th Symposium on Micro Nano Science and Technology in 2019.

Yuya Morimoto received his M.E. degree from the University of Tokyo in 2009. Between 2009 and 2011, he worked at Fujifilm Corporation on the R&D of medical endoscopes. He received his Ph.D. in mechano-informatics from the University of Tokyo in 2014. He is currently an assistant professor at the Institute of Industrial Science (IIS), University of Tokyo. His research interests are biohybrid robotics and biofabrication with microengineering techniques. He was a committee member of JSME 8th and 9th symposia on Micro Nano Science and Technology in 2017 and 2018.

micromachines

MDPI

Editorial

Editorial for the Special Issue of Selected Papers from the 8th Symposium on Micro–Nano Science and Technology on Micromachines

Norihisa Miki [1,*]**, Koji Miyazaki** [2] **and Yuya Morimoto** [3]

[1] Department of Mechanical Engineering, Keio University, 3-14-1 Hiyoshi, Kohoku-ku, Yokohama, Kanagawa 223-8522, Japan
[2] Department of Mechanical and Control Engineering, Kyushu Institute of Technology, 1-1 Sensui-cho, Tobata-ku, Kitakyushu, Fukuoka 804-8550, Japan; miyazaki@mech.kyutech.ac.jp
[3] Institute of Industrial Science (IIS), The University of Tokyo, 4-6-1 Komaba, Meguro-ku, Tokyo 153-8505, Japan; y-morimo@iis.u-tokyo.ac.jp
*   Correspondence: miki@mech.keio.ac.jp; Tel.: +81-45-566-1430

Received: 23 November 2018; Accepted: 25 November 2018; Published: 28 November 2018

The Micro–Nano Science and Technology Division of JSME (Japan Society of Mechanical Engineers) promotes academic activities to pioneer novel research topics on microscopic mechanics. The division encourages interdisciplinary studies to more deeply understand physical/chemical/biological phenomena on the micro/nano scale and to develop applied technologies. Since 2009, seven symposiums on Micro–Nano Science and Technology have taken place in a more interdisciplinary manner, incorporating the related societies of electronics and applied physics. We have promoted in-depth studies and interactions between researchers/engineers in various fields with more than 140 papers presented at each symposium over the past years. Thanks to these previous activities and the great effort of the committee members, the Micro–Nano Science and Technology Division has been recognized as a formal division of JSME.

This Special Issue collects 13 papers from the 8th Symposium on Micro–Nano Science and Technology, which was held from 31 October to 2 November, 2017 in Hiroshima, Japan. All of the papers highlight new findings and technologies at the micro/nano scales relating to a wide variety of fields of mechanical engineering, from fundamentals to applications.

Micro/nano fluidics have been studied using both fundamental and application-driven approaches. The visualization of the pH distribution around an ion depletion zone in a microchannel was successfully presented [1]. This technique and the knowledge that can be obtained by it will be indispensable for designing effective nano-channels for bio/chemical applications. Microfibers that can encapsulate cells and microbes will be a useful tool for fundamental biology, such as cell characterization and biomedical and environmental applications. The formation of branched and chained alginate microfibers was presented [2]. Considering the practical applications of bioremediation, a triple-coaxial flow device for the mass-production of hydrogel micro tubes containing microbes was designed and fabricated [3].

Materials and manufacturing technologies have always formed the core of micro/nano science and technologies. The crack-configuration of metal conductive tracks embedded in stretchable elastomers was analyzed thoroughly in [4], contributing to flexible electronics. The 3D shape reconstruction of 3D-printed transparent microscopic objects was demonstrated to further expand the design spaces for micro/nano structures [5]. The reductive sintering of mixed CuO/NiO nanoparticles using a femtosecond laser was intensively characterized, particularly with respect to heat accumulation [6]. The formation of arbitrary 3D shapes is a great challenge for micro/nano objects. Origami-like folding deformation [7] and self-organization using cellular automata [8] were successfully demonstrated.

Micromachines **2018**, *9*, 627

Physical sensors are a major application of micro/nano technologies. A biomimetic, artificial, sensory epithelium was designed and demonstrated [9]. Micro cantilever arrays work as the tactile sensor for gripping control in [10]. Simple methods for the reduction of the parasitic capacitance of a flexible polymer-based capacitive senor was proposed in [11].

The applications enabled by micro/nano technology-based sensors include human fatigue assessment [12] and adenosine triphosphate measurement in deep seas [13]. Such research requires the integration of the technologies of sensors, systems, and experiments. Micro/nano research mainly focuses on the phenomena and technologies at micro/nano scales. However, in order to make micro/nano research applicable, macroscopic viewing and technologies must be incorporated.

We would like to thank all the contributing authors for their excellent research work. We appreciate all the reviewers who provided valuable comments to improve the quality of the papers and the tremendous support from the editorial staff of *Micromachines*.

References

1. Mogi, K. A Visualization Technique of a Unique pH Distribution around an Ion Depletion Zone in a Microchannel by Using a Dual-Excitation Ratiometric Method. *Micromachines* **2018**, *9*, 167. [CrossRef] [PubMed]
2. Nishimura, K.; Morimoto, Y.; Mori, N.; Takeuchi, S. Formation of Branched and Chained Alginate Microfibers Using Theta-Glass Capillaries. *Micromachines* **2018**, *9*, 303. [CrossRef] [PubMed]
3. Fujimoto, K.; Higashi, K.; Onoe, H.; Miki, N. Development of a Triple-Coaxial Flow Device for Fabricating a Hydrogel Microtube and Its Application to Bioremediation. *Micromachines* **2018**, *9*, 76. [CrossRef] [PubMed]
4. Koshi, T.; Iwase, E. Crack-Configuration Analysis of Metal Conductive Track Embedded in Stretchable Elastomer. *Micromachines* **2018**, *9*, 130. [CrossRef] [PubMed]
5. Koyama, K.; Takakura, M.; Furukawa, T.; Maruo, S. 3D Shape Reconstruction of 3D Printed Transparent Microscopic Objects from Multiple Photographic Images Using Ultraviolet Illumination. *Micromachines* **2018**, *9*, 261. [CrossRef] [PubMed]
6. Mizoshiri, M.; Nishitani, K.; Hata, S. Effect of Heat Accumulation on Femtosecond Laser Reductive Sintering of Mixed CuO/NiO Nanoparticles. *Micromachines* **2018**, *9*, 264. [CrossRef] [PubMed]
7. Fukuie, K.; Iwata, Y.; Iwase, E. Design of Substrate Stretchability Using Origami-Like Folding Deformation for Flexible Thermoelectric Generator. *Micromachines* **2018**, *9*, 315. [CrossRef] [PubMed]
8. Ishida, T. Possibility of Controlling Self-Organized Patterns with Totalistic Cellular Automata Consisting of Both Rules like Game of Life and Rules Producing Turing Patterns. *Micromachines* **2018**, *9*, 339. [CrossRef] [PubMed]
9. Tsuji, T.; Nakayama, A.; Yamazaki, H.; Kawano, S. Artificial Cochlear Sensory Epithelium with Functions of Outer Hair Cells Mimicked Using Feedback Electrical Stimuli. *Micromachines* **2018**, *9*, 273. [CrossRef] [PubMed]
10. Araki, R.; Abe, T.; Noma, H.; Sohgawa, M. Miniaturization and High-Density Arrangement of Microcantilevers in Proximity and Tactile Sensor for Dexterous Gripping Control. *Micromachines* **2018**, *9*, 301. [CrossRef] [PubMed]
11. Nagatomo, T.; Miki, N. Reduction of Parasitic Capacitance of A PDMS Capacitive Force Sensor. *Micromachines* **2018**, *9*, 570. [CrossRef]
12. Horiuchi, R.; Ogasawara, T.; Miki, N. Fatigue Assessment by Blink Detected with Attachable Optical Sensors of Dye-Sensitized Photovoltaic Cells. *Micromachines* **2018**, *9*, 310. [CrossRef] [PubMed]
13. Fukuba, T.; Noguchi, T.; Okamura, K.; Fujii, T. Adenosine Triphosphate Measurement in Deep Sea Using a Microfluidic Device. *Micromachines* **2018**, *9*, 370. [CrossRef] [PubMed]

micromachines

MDPI

Article

A Visualization Technique of a Unique pH Distribution around an Ion Depletion Zone in a Microchannel by Using a Dual-Excitation Ratiometric Method

Katsuo Mogi

Molecular Profiling Research Center for Drug Discovery (Molprof), National Institute of Advanced Industrial Science and Technology (AIST), 2-4-7 Aomi, Koto-ku, Tokyo 135-0064, Japan; mogi.k@aist.go.jp; Tel.: +81-3-3599-8251

Received: 8 March 2018; Accepted: 27 March 2018; Published: 2 April 2018

Abstract: The ion depletion zone of ion concentration polarization has a strong potential to act as an immaterial barrier, separating delicate submicron substances, including biomolecules, without causing physical damage. However, the detailed mechanisms of the barrier effect remain incompletely understood because it is difficult to visualize the linked behavior of protons, cations, anions, and charged molecules in the thin ion depletion zone. In this study, pH distribution in an ion depletion zone was measured to estimate the role of proton behavior. This was done in order to use it as a tool with good controllability for biomolecule handling in the future. As a result, a unique pH peak was observed at several micrometers distance from the microchannel wall. The position of the peak appeared to be in agreement with the boundary of the ion depletion zone. From this agreement, it is expected that the pH peak has a causal connection to the barrier effect of the ion depletion zone.

Keywords: ion concentration polarization; ion depletion zone; Nafion; microfluidic device; pH indicator; fluorescein isothiocyanate (FITC)

---

1. Introduction

Understanding the ionic characteristics of molecules in solution is a useful tool for their handling, such as in concentration and separation. This is important in the fields of molecular chemistry and molecular biology, among others [1–4].

Ion concentration polarization (ICP), which can be easily used by applying voltage to a solution across an ion-exchange membrane, is a well-known and convenient concentration technique for dialysis [5–7]. Recently, not only this concentration technique, but another technique based on an almost negligible phenomenon called ion depletion, which occurs near a membrane under steady-state ICP, have attracted attention [8,9]. The microscale ion depletion zone has a peculiar ability to repel charged particles, including biomolecules. The strong potential of using this effect to provide an immaterial barrier for the separation of delicate submicron substances, without causing physical damage, has been reported [10–12]. The microscale barrier can be effectively employed by building an ICP system into a microfluidic device, in which a microchannel provides sufficient space for using the valid range of the barrier [13].

It is generally believed that the microscale barrier effect of the ion depletion zone is caused by nanoscale ionic behavior. This is due to the fluidic and electrochemical forces around the ion-exchange membrane [9]. Hence, it is important to investigate pH distribution in nanochannels and porous materials [14–17]. However, the detailed mechanism of the barrier effect remains incompletely understood because it is difficult to visualize the linked behavior of protons, cations, anions, and charged molecules in the thin ion depletion zone.

In this study, to make better use of the barrier effect by controlling the volume of ion transfer, including protons, the proton concentration distribution in the ion depletion zone was verified to clarify the role of proton behavior in the barrier effect. For verification, the pH was measured by the dual excitation ratio method using fluorescein isothiocyanate (FITC), one of the most common negatively charged fluorophores used as a pH indicator [18,19]. Furthermore, the boundary of the barrier of the ion depletion zone was estimated from the concentration distribution of the FITC molecule, which was subjected to a repulsive force.

2. Materials and Methods

2.1. Ion Concentration Polarization (ICP) in Microchannels

To make effective use of the almost negligible phenomena in an ion depletion zone, an ICP system was scaled down using a microfluidic device [20,21]. Figure 1 shows an illustration of two microchannels in the microfluidic device. These two microchannels were connected by a cation-exchange membrane. The cross-sectional view of A-A' shows the movement of cations and anions in the channels. When voltage was applied to generate a difference in potential between the two channels, with the right channel at a higher potential than the left, cations moved through the membrane from right to left and anions remained in their original channels. This formed an ion depletion zone around the right side of the membrane under steady-state ICP, and this ion depletion zone repelled all charged particles. While the ion depletion zone expanded due to the potential difference, the dimension of the zone was maintained by ions which were continuously supplied by constant flow.

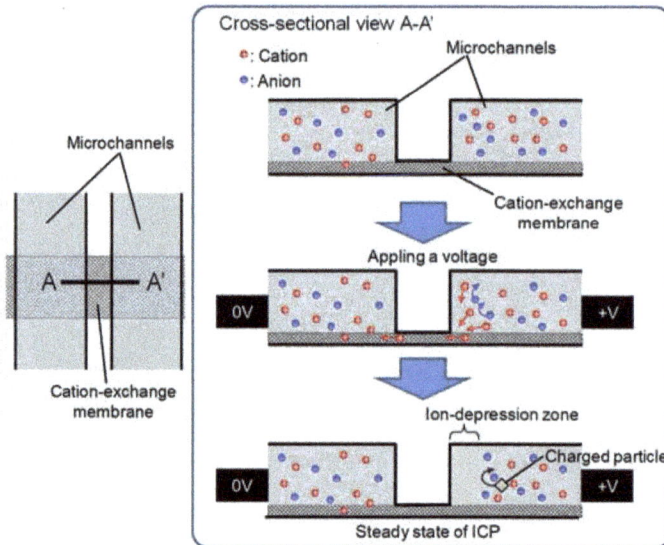

Figure 1. Schematic of microchannels for ion concentration polarization (ICP). The left schematic is a top view of a region composed of two microchannels and a cation-exchange membrane. The right schematic is a cross-sectional view of A-A' in the left schematic. Generating a difference in potential causes the movement of cations through the membrane, whereas anions remain in their original channels. Under steady-state ICP, an ion depletion zone is formed around the high-potential side of the membrane.

2.2. Measurement Principles

To visualize proton behavior as an indicator of ion balance, the pH was measured around the ion depletion zone. The pH in the microchannel was non-invasively measured by a dual-excitation ratiometric technique using FITC [22]. In a microchannel with a short light path for observation, the fluorescent intensity of FITC, I_f, excited by excitation beam of intensity I_e is given by Equation (1):

$$I_f(\text{pH}, C) = I_e C \phi_e \varepsilon_e(\text{pH}) - I_{back,e}(\text{pH}, C) \tag{1}$$

In Equation (1), e (nm) is the wavelength of the beam, C (kg/m^2) is the FITC concentration, ϕ_e is the quantum efficiency, ε_e (m^2/kg) is the molar absorption coefficient, and $I_{back,e}$ is the background intensity. Hence, the normalized fluorescent intensity depends only on the pH, as given by Equation (2):

$$I_f(\text{pH}) = \frac{I_{488}\phi_{488}\varepsilon_{488}(\text{pH}) - I_{back,488}(\text{pH})}{I_{458}]\phi_{458}\varepsilon_{458}(\text{pH}) - I_{back,458}(\text{pH})} \tag{2}$$

2.3. Standard Solutions for pH Calibration

Standard solutions for pH calibration containing 9×10^{-6} M FITC were adjusted to pH 2.05 and 2.68 with sodium dihydrogenphosphate dihydrate and phosphoric acid, to pH 4.61 with potassium dihydrogenphosphate, to pH 5.74, 6.52, and 7.28 with potassium dihydrogenphosphate and disodium hydrogen phosphate 12-water, to pH 6.76 with sodium dihydrogenphosphate dehydrate and disodium hydrogen phosphate 12-water, and to pH 8.18 with potassium dihydrogenphosphate and potassium hydrate.

3. Experimental Section

3.1. Experimental Setup

The microfluidic ICP device was composed of a glass substrate with a cation-exchange membrane pattern and a polydimethylsiloxane (PDMS) substrate with two microchannels, as shown in Figure 2a. One of the channels, which was 1 mm wide and 100 μm high, was used for the injection of FITC solutions and was called the "main channel". The other channel, used for the injection of distilled water, was named the "Ground (GND) channel" and was 2 mm wide and 100 μm high. The central areas of the microchannels were connected with a Nafion (DuPont) membrane, which is a band-shaped pattern with a width of 100 μm across the channels. The FITC solution in the main channel and the distilled water in the GND channel were drawn in at 5 μL/min by a syringe pump (KDS 200 syringe pump, KD Scientific Inc., Holliston, MA, USA). Then, ion transfer for ICP around the Nafion pattern in the main channel was generated by applying a voltage of 30 V. A voltage supply (P4K-80M, Matsusada Precision Inc., Bohemia, NY, USA) was connected to the electrode on the inlet and outlet ports of the microchannels.

The fluorescent intensity of FITC in a microchannel was measured with a confocal laser scanning microscope system (TCS STED-CW; Leica Microsystems, Leica Microsystems Inc., Buffalo Grove, IL, USA), as shown in Figure 2b [23]. Excitation beams at 458 nm and 488 nm were selected from an Ar-ion laser source using an acousto-optic tunable filter (AOTF). A 1.55 mm^2 field of the main channel around the Nafion was scanned with the excitation beam, which was controlled by galvanometer mirrors. The FITC fluorescence was obtained by avalanche photodiode as a 16-bit intensity image.

3.2. Device Fabrication

The microchannels were fabricated on PDMS using a commonly-employed soft lithography technique [24–26]. A band-shaped cation-exchange membrane was patterned on a glass substrate using a Nafion solution (Nafion®20 wt % dispersion, DuPont). The shape of the Nafion solution was formed using a PDMS mold with a microchannel 100 μm wide and 25 μm high. Uncured

Nafion solution was injected into the channel of the PDMS mold on the glass substrate, as shown in Figure 3. Then, the substrate with the PDMS mold containing the Nafion solution was cured at 100 °C for 10 min. After peeling off the mold, the Nafion pattern was formed on the glass substrate. The Nafion-patterned glass substrate was aligned with the microchannel substrate made of PDMS to complete the microfluidic ICP device.

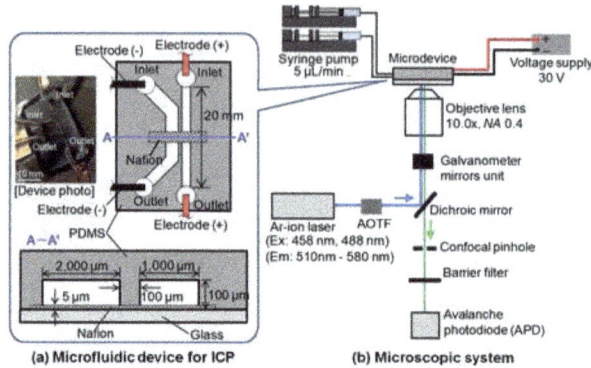

Figure 2. Schematic of the experimental system. (**a**) Microfluidic device for ICP made of glass and substrates; (**b**) A schematic of the microscopic system used for pH measurement. Excitation beams at 458 nm and 488 nm were used in a dual-excitation ratiometric method. The microfluidic device on the microscopic system was connected to a syringe pump and a voltage supply.

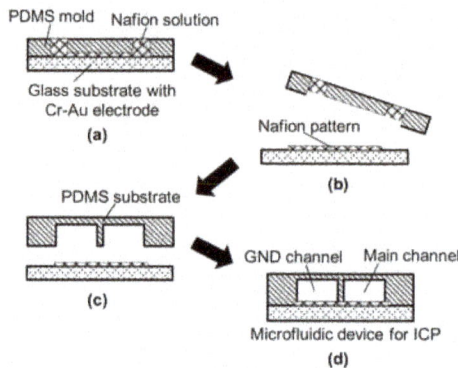

Figure 3. Schematic of the fabrication process for Nafion patterning. (**a**) Cross-sectional view of Nafion patterned on a glass substrate using a polydimethylsiloxane (PDMS) mold; (**b**) Nafion pattern cured at 100 °C for 10 min after peeling off the mold; (**c**) Assembly of the PDMS substrate onto the glass substrate; (**d**) Completed microfluidic ICP device.

4. Results and Discussion

To generate a calibration curve of pH vs. the experimental value, fluorescent intensities obtained from pH standard solutions with pH values of 2.05, 2.68, 4.61, 5.74, 6.52, 6.76, 7.28, and 8.18 were measured by dual excitation at 458 nm and 488 nm, as shown in Figure 4a. Each intensity value in the figure is the mean value over the scanned area (28×28 μm^2 square) at the center of the microchannel, and the error bar shows the standard deviation of the pixel data. From the intensity values in Figure 4a, the quotient of the intensities represented by Equation (2), normalized by the maximum value, was obtained as shown in Figure 4b. The calibration curve in Figure 4b was generated by a polynomial approximation, and the R-squared value was 1.0 in the range from pH = 2.68 to pH = 7.28. Hence, it was decided that the calibration curve is useful to estimate the pH distribution in ranges other than pH = 2.05 and pH = 8.18. Therefore, it was decided that this calibration curve would be applied to the estimation of the pH distribution in that range.

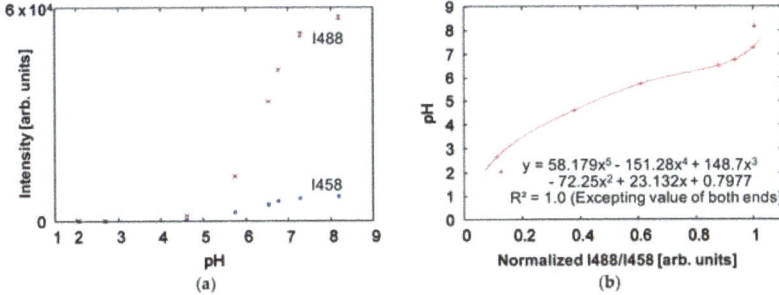

Figure 4. Calibration curve between pH value and fluorescein isothiocyanate (FITC) intensity. (**a**) FITC intensities of pH standard solutions obtained by dual excitation at 458 nm and 488 nm. Each intensity shown is the mean value over a 28×28 μm^2 square, and the error bar shows the standard deviation; (**b**) A calibration curve between pH value and the quotient of the intensities, normalized by the maximum value. The calibration curve is a polynomial approximation: $y = 58.2x^5 - 151.3x^4 + 148.7x^3 - 72.3x^2 + 23.1x + 0.8$. The R-squared value was 1.0 from a pH of 2.68 to a pH of 7.28, excluding the endpoints of 2.05 and 8.18.

To estimate the pH distribution in the ion depletion zone based on the calibration curve, an image of the fluorescent intensity of FITC around the Nafion pattern was captured, as shown in Figure 5a. The zero position of the x-axis was defined as the wall surface of the main channel, and the zero position of the y-axis was defined as the center of the Nafion pattern. Figure 5b shows an intensity distribution in the steady stage at lines a, b and c in Figure 5a. The lines are parallel with the x-axis, and the positions on the y-axis are y = 139 μm, 0 μm and -139 μm. The effect of the autofluorescence of the Nafion was removed by normalization using the background intensity of the channel without Nafion.

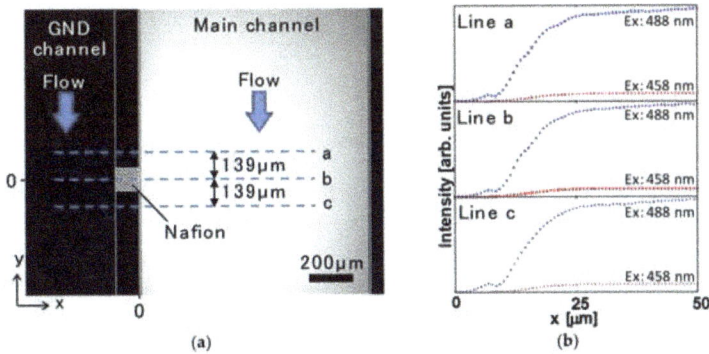

Figure 5. Measurement area and the FITC intensity distribution. (**a**) Photo of fluorescent intensity of FITC in steady state at the ion depletion zone in the microchannel. The zero position of the x-axis is the wall surface of the main channel, and the zero position of the y-axis is the center of the Nafion pattern. Lines a, b and c are parallel to the x-axis, and their positions on the y-axis are y = 139 μm, 0 μm and -139 μm; (**b**) Intensity distributions obtained by dual excitation at 458 nm and 488 nm under a steady-state ion depletion zone at lines a, b and c.

Figure 6a shows the FITC concentration distributions at the lines a, b and c in Figure 5a, as obtained from the fluorescent intensity excited by the 458 nm laser. As seen in the figure, the rise curve of the concentration drifted from the microchannel wall to the center of the channel by the repulsive force of the ion depletion zone, and the top of the concentration curve became 1.5 times higher from the force. In this study, the rising point ($x = 9.0 \pm 0.10$ μm) of the drifted curve, which was the x-intercept of the approximate line between $x = 11$ μm and 19 μm, was defined as the boundary of the ion depletion zone.

On the other hand, Figure 6b shows the value of the pH distribution ("I488/I458") obtained from the intensity quotient values for excitation at 458 nm and 488 nm. Scrupulous attention is required to treat the pH values, because the values lower than pH 2.68 and higher than pH 7.28 were out of range in application of the calibration curve. As illustrated by the dotted line in Figure 6b, a convex peak appeared while applying ICP. The position coordinate of the peak is $x = 6.3 \pm 0.15$ μm, which was obtained by Gaussian fitting. Although measurement error is considered to be included due to the low concentration of FITC at the right of the convex peak, a slight increase in pH due to a decrease in proton by passing through the Nafion pattern was observed at the closest side of the microchannel wall. Interestingly, it can be clearly seen that the bottom of the concave peak ($x = 9.1 \pm 0.08$ μm) to the right of the convex peak corresponded well with the boundary of the ion depletion zone, as shown in Figure 6c. It may be suspected that the sharp increase and decrease in pH is a factor in forming a specific electrochemical equilibrium state that generates the boundary of an ion depletion zone, which acts as a barrier to keep charged substances away.

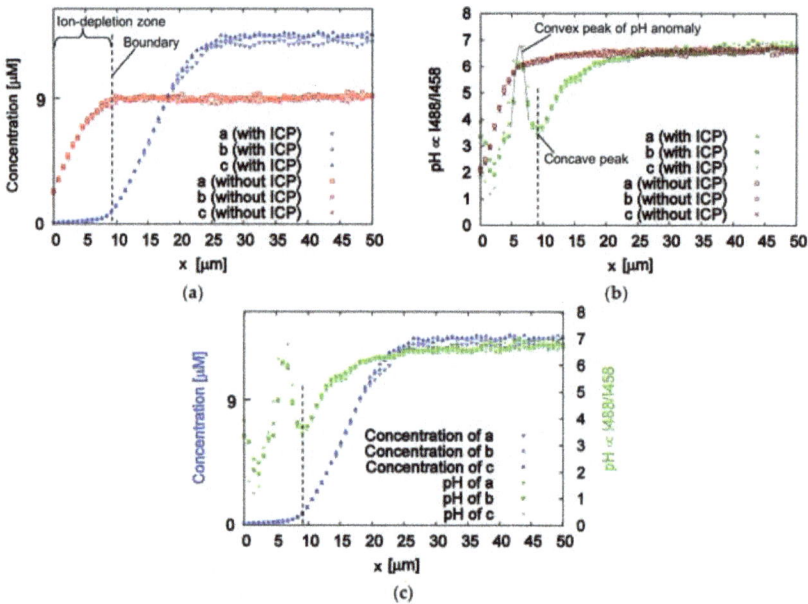

Figure 6. FITC concentration distribution and pH distribution at the ion depletion zone generated by ICP. (**a**) FITC concentration distributions under steady state at lines a, b and c while applying ICP, with ICP and without ICP. The dotted line at $x = 9.0 \pm 0.10$ μm is the boundary of the ion depletion zone; (**b**) pH distribution obtained from the intensity quotient values after excitation at 458 nm and 488 nm ("I488/I458"), with ICP and without ICP. The pH distribution with ICP has a convex peak at $x = 6.3 \pm 0.15$ μm and a concave peak at $x = 9.1 \pm 0.08$ μm; (**c**) FITC concentration and pH distributions at lines a, b, and c under steady-state ICP.

5. Conclusions

The pH distribution around an ion depletion zone in a microchannel was measured by a dual excitation ratio method with FITC to estimate proton behavior. In a microchannel of PDMS without ICP, pH is slightly decreased near the microchannel wall due to the electric double layer. On the other hand, a unique pH peak, which has never been previously reported, was observed at $x = 6.3 \pm 0.15$ μm from the microchannel wall in cases with ICP. Furthermore, the position of the unique peak was in agreement with the boundary of the ion depletion zone, which was estimated from the rising point of the FITC concentration. This agreement indicates that the barrier effect of the ion depletion zone has a profound causal connection with the pH anomaly. Although there has not yet been enough investigation to clarify the phenomena related to ionic behavior in the ion depletion zone, it can be expected that this report may play an important role in better utilizing barrier effects with high controllability for biomolecule handling in the future.

Acknowledgments: This work was partially supported by the Ministry of Education, Science, Sports and Culture, Japan, a grant-in-aid for young scientists (B) (No. 15K20989 and 17K17715) and JKA RING!RING! Project for the financial support (FY2015, No. 27-183). We would like to thank the 4 University Nano Micro Fabrication Consortium in Kawasaki, Japan (http://www.nano-micro.sakura.ne.jp/home/) who provided open facilities and experimental equipment for this work.

Conflicts of Interest: The author declares no conflict of interest.

References

1. Medintz, I.L.; Paegel, B.M.; Mathies, R.A. Microfabricated capillary array electrophoresis DNA analysis systems. *J. Chromatogr. A* **2001**, *924*, 265–270. [CrossRef]
2. Chou, H.P.; Spence, C.; Scherer, A.; Quake, S. A microfabricated device for sizing and sorting DNA molecules. *Proc. Natl. Acad. Sci. USA* **1999**, *96*, 11–13. [CrossRef] [PubMed]
3. Fiedler, S.; Shirley, S.G.; Schnelle, T.; Fuhr, G. Dielectrophoretic sorting of particles and cells in a microsystem. *Anal. Chem.* **1998**, *70*, 1909–1915. [CrossRef] [PubMed]
4. Shields, C.W.; Reyes, C.D.; Lopez, G.P. Microfluidic cell sorting: A review of the advances in the separation of cells from debulking to rare cell isolation. *Lab Chip* **2015**, *15*, 1230–1249. [CrossRef] [PubMed]
5. Xu, T.W.; Yang, W.H. Sulfuric acid recovery from titanium white (pigment) waste liquor using diffusion dialysis with a new series of anion exchange membranes—Static runs. *J. Membr. Sci.* **2001**, *183*, 193–200.
6. Bhattacharjee, S.; Chen, J.C.; Elimelech, M. Coupled model of concentration polarization and pore transport in crossflow nanofiltration. *AIChE J.* **2001**, *47*, 2733–2745. [CrossRef]
7. Kim, J.; Cho, I.; Lee, H.; Kim, S.J. Ion concentration polarization by bifurcated current path. *Sci. Rep.* **2017**, *7*, 5091. [CrossRef] [PubMed]
8. Jeon, H.; Lee, H.; Kang, K.H.; Lim, G. Ion concentration polarization-based continuous separation device using electrical repulsion in the depletion region. *Sci. Rep.* **2013**, *3*, 3483. [CrossRef] [PubMed]
9. Kim, S.J.; Ko, S.H.; Kang, K.H.; Han, J. Direct seawater desalination by ion concentration polarization. *Nat. Nanotechnol.* **2010**, *5*, 297–301. [CrossRef] [PubMed]
10. Hyashida, K.; Mogi, K.; Yamamoto, T. Development of separation-and-concentration device for micro/nano particles by ion concentration polarization. In Proceedings of the 6th JSME Micro-Nano Symposium, Matsue, Shimane, Japan, 20–22 October 2014.
11. Hyashida, K.; Mogi, K.; Honda, A.; Yamamoto, T. Separation and condensation of bio-nanoparticles using ion depletion effect. In Proceedings of the 7th JSME Micro-Nano Symposium, Niigata, Japan, 28–30 October 2015.
12. Kwak, R.; Kim, S.J.; Han, J. Continuous-flow biomolecule and cell concentrator by ion concentration polarization. *Anal. Chem.* **2011**, *83*, 7348–7355. [CrossRef] [PubMed]
13. Mark, D.; Haeberle, S.; Roth, G.; von Stetten, F.; Zengerle, R. Microfluidic lab-on-a-chip platforms: Requirements, characteristics and applications. *Chem. Soc. Rev.* **2010**, *39*, 1153–1182. [CrossRef] [PubMed]
14. Chang, C.C.; Yeh, C.P.; Yang, R.J. Ion concentration polarization near microchannel-nanochannel interfaces: Effect of pH value. *Electrophoresis* **2012**, *33*, 758–764. [CrossRef] [PubMed]
15. Fuest, M.; Rangharajan, K.K.; Boone, C.; Conlisk, A.T.; Prakash, S. Cation dependent surface charge regulation in gated nanofluidic devices. *Anal. Chem.* **2017**, *89*, 1593–1601. [CrossRef] [PubMed]

16. Su, J.; Guo, H. Effect of nanochannel dimension on the transport of water molecules. *J. Phys. Chem. B* **2012**, *116*, 5925–5932. [CrossRef] [PubMed]

17. Yeh, L.H.; Zhang, M.; Qian, S. Ion transport in a pH-regulated nanopore. *Anal. Chem.* **2013**, *85*, 7527–7534. [CrossRef] [PubMed]

18. Han, J.Y.; Burgess, K. Fluorescent indicators for intracellular pH. *Chem. Rev.* **2010**, *110*, 2709–2728. [CrossRef] [PubMed]

19. Martin, M.M.; Lindqvist, L. PH-dependence of fluorescein fluorescence. *J. Lumin.* **1975**, *10*, 381–390. [CrossRef]

20. Mogi, K.; Sugii, Y.; Yamamoto, T.; Fujii, T. Rapid fabrication technique of nano/microfluidic device with high mechanical stability utilizing two-step soft lithography. *Sens. Actuators B Chem.* **2014**, *201*, 407–412. [CrossRef]

21. Mogi, K.; Sugii, Y.; Fujii, T.; Matsumoto, Y. A microfluidic culturing system for observation of free-floating microorganisms. In Proceedings of the 11th International Conference on Nanochannels, Microchannels, and Minichannels, Sapporo, Japan, 16–19 June 2013.

22. Ichiyanagi, M.; Sato, Y.; Hishida, K. Optically sliced measurement of velocity and pH distribution in microchannel. *Exp. Fluids* **2007**, *43*, 425–435. [CrossRef]

23. Kazoe, Y.; Mawatari, K.; Sugii, Y.; Kitamori, T. Development of a measurement technique for ion distribution in an extended nanochannel by super-resolution-laser-induced fluorescence. *Anal. Chem.* **2011**, *83*, 8152–8157. [CrossRef] [PubMed]

24. Xia, Y.N.; Whitesides, G.M. Soft lithography. *Angew. Chem. Int. Ed.* **1998**, *37*, 550–575. [CrossRef]

25. Whitesides, G.M.; Ostuni, E.; Takayama, S.; Jiang, X.Y.; Ingber, D.E. Soft lithography in biology and biochemistry. *Annu. Rev. Biomed. Eng.* **2001**, *3*, 335–373. [CrossRef] [PubMed]

26. Mogi, K.; Fujii, T. A novel assembly technique with semi-automatic alignment for pdms substrates. *Lab Chip* **2013**, *13*, 1044–1047. [CrossRef] [PubMed]

micromachines

MDPI

Article

Formation of Branched and Chained Alginate Microfibers Using Theta-Glass Capillaries

Keigo Nishimura [1] , Yuya Morimoto [1] , Nobuhito Mori [1,†] and Shoji Takeuchi [1,2,*]

1   Center for International Research on Integrative Biomedical Systems (CIBiS), Institute of Industrial Science (IIS), The University of Tokyo, 4-6-1 Komaba, Meguro-ku, Tokyo 153-8505, Japan; nishim@iis.u-tokyo.ac.jp (K.N.); y-morimo@iis.u-tokyo.ac.jp (Y.M.); mori1985@iis.u-tokyo.ac.jp (N.M.)
2   International Research Center for Neurointelligence (WPI-IRCN), The University of Tokyo Institutes for Advanced Study (UTIAS), The University of Tokyo, 4-6-1 Komaba, Meguro-ku, Tokyo 153-8505, Japan
*   Correspondence: takeuchi@iis.u-tokyo.ac.jp; Tel.: +81-3-5452-6650; Fax: +81-3-5452-6649
†   Current address: Biotechnology Research Institute for Drug Discovery, National Institute of Advanced Industrial Science and Technology (AIST), 1-1-1 Umezono, Tsukuba, Ibaraki 305-8560, Japan.

Received: 10 May 2018; Accepted: 14 June 2018; Published: 17 June 2018

Abstract: This study proposes a microfluidic spinning method to form alginate microfibers with branched and chained structures by controlling two streams of a sodium alginate solution extruded from a theta-glass capillary (a double-compartmented glass capillary). The two streams have three flow regimes: (i) a combined flow regime (single-threaded stream), (ii) a separated flow regime (double-threaded stream), and (iii) a chained flow regime (stream of repeating single- and double-threaded streams). The flow rate of the sodium alginate solution and the tip diameter of the theta-glass capillary are the two parameters which decide the flow regime. By controlling the two parameters, we form branched (a Y-shaped structure composed of thick parent fiber and permanently divided two thin fibers) and chained (a repeating structure of single- and double-threaded fibers with constant frequency) alginate microfibers with various dimensions. Furthermore, we demonstrate the applicability of the alginate microfibers as sacrificial templates for the formation of chain-shaped microchannels with two inlets. Such microchannels could mimic the structure of blood vessels and are applicable for the research fields of fluidics including hemodynamics.

Keywords: microfluidics; microfiber spinning; alginate hydrogel

1. Introduction

Alginate microfibers have been attractive materials for biomedical applications such as cell and drug encapsulation [1–3], scaffolds for cell culture [4–6], and channel formation as sacrificial templates [7–9] because of their biocompatibility, biodegradability, and mechanical flexibility [1]. There have been two main methods to form the alginate microfibers: electrospinning and microfluidic spinning [10–12]. On one hand, electrospinning is suitable for forming high-resolution micro- and nano-scaled fibers, but the setup needs high voltages and there is a material limitation that alginate solutions cannot be independently electrospun without mixing them with other polymers such as chitosan [13,14]. On the other hand, microfluidic spinning, which involves a continuous extrusion of sodium alginate solution into a calcium chloride solution bath via microfluidic devices, has been widely used because of its easy setup and the adjustability of dimensions of microfibers by modulating flow conditions such as flow rates. Though resolutions of microfluidic spinning are lower than electrospinning, microfluidic spinning can form microfibers with various cross-sectional shapes. The alginate microfibers formed by microfluidic spinning have been shown to achieve not only uniform cross-sectional shapes when the extrusion of sodium alginate solution is constant (e.g., fiber [15–17], tube [18,19], and fiber with grooved surface [20,21]), but also cross-sectional shapes have been shown to change continuously when the extrusion is dynamically varied (e.g., coded fiber [20] and beads-on-a-string [22]). However, conventional microfluidic spinning methods can only generate a single-threaded microfiber, thus failing to form more complicated geometries such as in branched or chained microfibers.

In this study, we present a microfluidic spinning method for the formation of branched/chained alginate microfibers by controlling the flow regime of a sodium alginate solution extruded from a theta-glass capillary (Figure 1a). This capillary has double compartments composed of a center partition and two parallel channels. In our experiment, we found that the two streams extruded from the capillary show three flow regimes: (i) a combined flow (single-threaded stream), (ii) a separated flow (double-threaded stream), and (iii) a chained flow (stream of repeating single- and double-threaded streams) based on the flow rate of the sodium alginate solution and the tip diameter of the theta-glass capillary (Figure 1b). Here, by controlling the flow regime, we try to fabricate alginate microfibers with branched/chained structures having various dimensions. We first reveal the relationship between the flow regimes and flow condition including the flow rate of the sodium alginate solution and the tip diameter of the theta-glass capillary, and then apply the method to form alginate microfibers with branched and chained structures. As a demonstration of the microfibers' applicability, we form microchannels using the microfibers with branched structures as a sacrificial template.

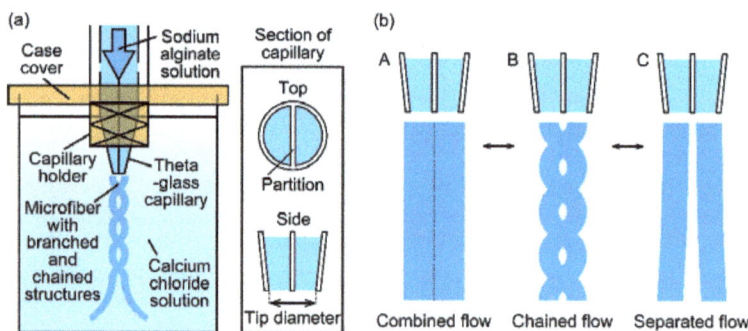

Figure 1. Schematic of the formation of alginate microfibers: (**a**) formation of alginate microfibers with branched structures using a theta-glass capillary; (**b**) conceptual illustrations of three flow regimes of two streams extruded from the theta-glass capillary (A: combined flow, B: chained flow, C: separated flow).

2. Materials and Methods

2.1. Materials

To form alginate microfibers, we prepared a sodium alginate solution and 150-mM calcium chloride solution by dissolving sodium alginate powder (Junsei Chemical Co., Ltd., Tokyo, Japan) and calcium chloride powder (Kanto Chemical Co., Inc., Tokyo, Japan) in deionized water, respectively. To visualize the sodium alginate solution under a bright field or fluorescent microscopy, we added 5% (v/v) blue ink (PILOT Corporation, Tokyo, Japan) or 0.04% (w/v) fluorescent nanobeads (FluoSpheres™ carboxylate-modified microspheres, 0.2 μm, yellow-green fluorescent (505/515), 2% solids, Life Technologies Corp., Carlsbad, CA, USA) to the solution. As materials for microchannels, we used a polydimethylsiloxane (PDMS) elastomer and a curing agent (Silpot 184 W/C, Dow Corning Toray Co., Ltd., Tokyo, Japan). To form PDMS microchannels, we prepared a 500-mM sodium citrate solution as a chelating agent to wash out alginate microfibers by dissolving sodium citrate (Nacalai Tesque, Inc., Kyoto, Japan) in deionized water. To demonstrate a microfluidic operation in the

PDMS microchannels, we prepared 5% (*v/v*) red ink (PILOT Corporation, Tokyo, Japan) and green ink (Pelikan Vertriebsgesellschaft mbH & Co. KG, Hannover, Germany) by diluting them with deionized water.

2.2. Device Fabrication

The device used to form alginate microfibers was composed of a theta-glass capillary, capillary holder, and case cover. The tip of the theta-glass capillary (TST150-6, World Precision Instruments, Sarasota, FL, USA) was sharpened with a puller (PC-10, Narishige Co., Ltd., Tokyo, Japan). To adjust the tip diameters (tolerance: ±10 μm), we cut the sharpened tip using a microforge (MF-900, Narishige Co. Ltd., Tokyo, Japan) and grinded it using a microgrinder (EG-400, Narishige Co., Ltd., Tokyo, Japan) (Figure 2a). A capillary holder and case cover for immobilizing the theta-glass capillary were fabricated using a 3D printer (AGILISTA-3100, KEYENCE Corporation, Osaka, Japan).

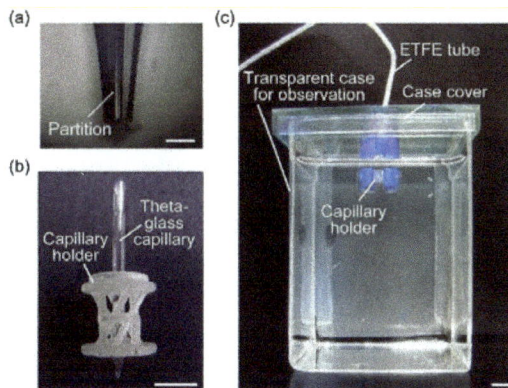

Figure 2. Experimental setups for alginate microfiber formation: (**a**) sharpened theta-glass capillary; (**b**) capillary holder assembled with the sharpened theta-glass capillary; (**c**) experimental setup for formation of alginate microfibers. Scale bars are: (**a**) 200 μm and (**b,c**) 5 mm.

2.3. Formation of Alginate Microfibers

A 5-mL syringe (Terumo Corp., Tokyo, Japan) containing a sodium alginate solution was connected to the theta-glass capillary with silicone tubes (AS ONE Corporation, Osaka, Japan) and ethylene tetrafluoroethylene (ETFE) tubes (VICI Precision Sampling, Inc., Baton Rouge, LA, USA). The theta-glass capillary was vertically set to the capillary holder and case cover and then placed on a clear case filled with a calcium chloride solution (Figure 2b,c). The theta-glass capillary tip was submerged under the calcium chloride solution, and the direction of the capillary outlet was approximately aligned to the direction of gravitational force. When the sodium alginate solution was extruded into the calcium chloride solution using a syringe pump (KDS 210, KD Scientific Inc., Holliston, MA, USA) under precise control of the flow rate, calcium-alginate hydrogel microfibers were formed by a crosslinking reaction of the sodium alginate solution with calcium ions. The formation of the microfibers was observed using a high-speed microscope (VW-9000, KEYENCE Corp., Osaka, Japan) and we classified the flow regimes according to the shapes of the alginate solution streams.

2.4. Fabrication of PDMS Microchannel Using Alginate Microfiber as Sacrificial Template

To fabricate chained PDMS microchannels, we used a modified method of hydrogel molding techniques [8]. To briefly describe the process, we first prepared an alginate microfiber with a structure where two long and thin fibers are connected to a chained fiber by controlling the flow rates of the sodium alginate solution in order to switch the flow regimes (i.e., from separated to chained). A thin

PDMS substrate was formed in a 60-mm cell culture dish (Corning Inc., Corning, NY, USA) by heating a PDMS-curing agent mixture at a ratio of 10 to 1 (*w/w*) on the dish bottom for 90 min at 75 °C. Next, the PDMS-curing agent mixture was poured onto the thin PDMS substrate with the alginate microfiber on its surface and cured by leaving it in a vacuum desiccator overnight at room temperature (Figure 3a). The PDMS structure with the alginate microfiber was removed from the dish and was punched with a biopsy punch (tip diameter 1.5 mm, Kai Industries Co., Ltd., Gifu, Japan) at the ends of the embedded microfiber to form one-side-open holes used as inlets and an outlet of a microchannel. To form inlets and an outlet without leakage, we inserted ETFE tubes in the holes and sealed the gap between the holes and tubes by forming a PDMS cover on the PDMS structure (Figure 3b). To form microchannels, we washed out the embedded microfiber by introducing 500-mM sodium citrate solution in the PDMS substrate through the inlets (Figure 3c). One of the fabricated PDMS microchannels (Figure 3d,e) was sliced into cross-sectional thin layers using disposable microtome knives (S35, FEATHER Safety Razor Co., Ltd., Osaka, Japan) in order to observe the cross sections. To demonstrate a microfluidic operation in the microchannel, 5% (*v/v*) red and green ink, which were loaded in 1-mL gastight syringes (Model 1001, Hamilton Co., Reno, NV, USA), were introduced using a syringe pump.

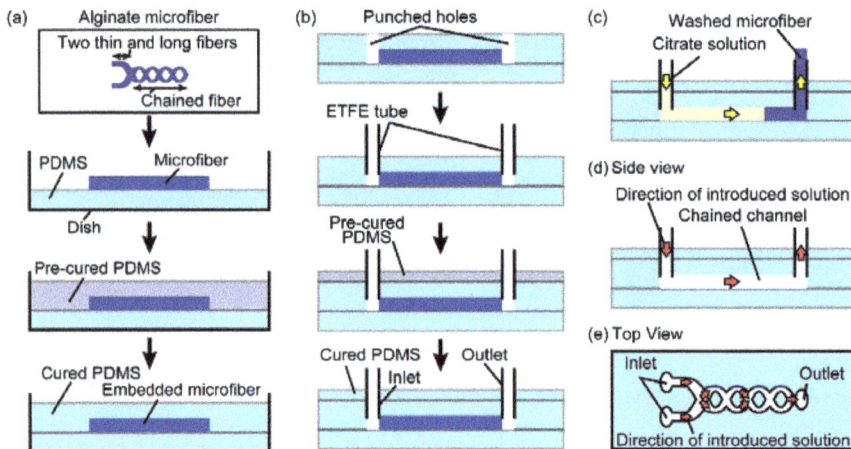

Figure 3. Process flow of the fabrication of chained PDMS microchannels. (**a**) Embedding of a microfiber with structure that two thin and long fibers are connected to a chained fiber into the PDMS substrate; (**b**) fabrication of inlets and an outlet for the microchannel; (**c**) removing the embedded microfiber; (**d**) side view of the formed microchannel; (**e**) top view of the formed microchannel.

3. Results and Discussion

3.1. Characterization of Flow Regimes of Sodium Alginate Solution

To investigate the variation of flow regimes of the sodium alginate solution extruded from the theta-glass capillary into the calcium chloride solution, we changed the flow rates of the sodium alginate solution to within a range of 0.01–15 mL/min. When we extruded 1.5% (*w/w*) sodium alginate solution into 150-mM calcium chloride solution through the theta-glass capillary with a 500-μm tip diameter, we observed that streams of the sodium alginate solution had various shapes depending on the flow rate (Figure 4, Movie S1). As we observed in the preliminary experiment, we confirmed that the streams can be classified into three flow regimes: combined flow regime (Figure 4a(I),e(II)), separated flow regime (Figure 4c(I),g(II)), and chained flow regime (Figure 4b(I),d(II),f(III)). In the combined flow regime occurring at 0.01 mL/min and 5.0–8.0 mL/min, a single-threaded stream was formed by integrating two streams at the capillary tip. In the separated flow regime occurring at 0.1–4.0 and 10–15 mL/min, a double-threaded stream was formed by retaining the two streams. By contrast, in the chained flow regime occurring at the narrow flow condition (0.02–0.05, 4.1, and 9.0 mL/min) between the combined and separated flow regimes, a chained stream was formed by repeated formations of single- and double-threaded streams. These results indicate that our method can generate streams of sodium alginate solution with various configurations.

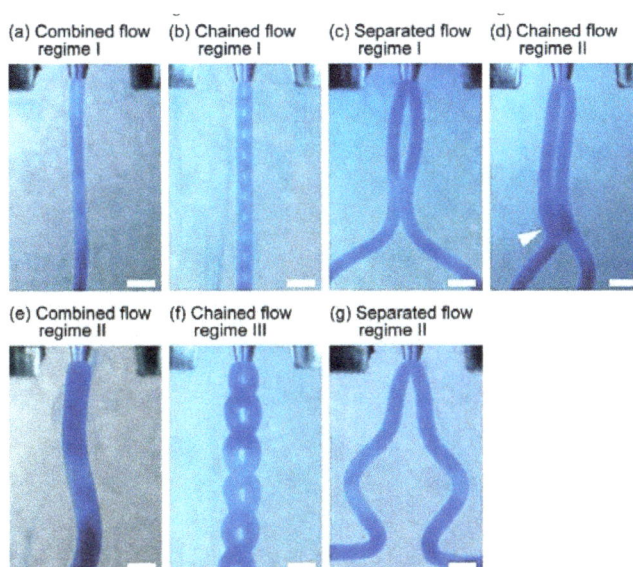

Figure 4. Flow regimes of a sodium alginate solution extruded into a calcium chloride solution at various flow rates through a theta-glass capillary: (**a**) combined flow regime I (0.01 mL/min); (**b**) chained flow regime I (0.03 mL/min); (**c**) separated flow regime I (3.0 mL/min); (**d**) chained flow regime II (4.1 mL/min). The arrowhead shows the combined part of the flow; (**e**) combined flow regime II (7.0 mL/min); (**f**) chained flow regime III (9.0 mL/min); (**g**) separated flow regime II (11 mL/min). Scale bars are 1 mm.

Moreover, we investigated the relationships among the flow regimes, the flow rate of the sodium alginate solution, and the tip diameter of the theta-glass capillary and produced an experimental phase diagram (Figure 5 and Figure S1). Under the conditions of low flow rates (0.01–4 mL/min) and low tip diameters (100 and 200 µm), middle flow rates (2.0–8.0 mL/min) and a middle tip diameter (300 µm), and high flow rates (4.0–15 mL/min) and large tip diameters (400 and 500 µm), results showed that flow regimes were arranged in an orderly manner and that all three kinds of flow regimes appeared at every tip diameter. These flow regimes corresponded to combined flow regime II, chained flow regime III, and separated flow regime II observed in the aforementioned results. By contrast, under the conditions of low flow rates (0.01–4.0 mL/min) and large tip diameters (300–500 µm), we found that all three kinds of flow regimes, whose streams had smaller diameters, were also arranged in an orderly manner when tip diameters were 300 and 500 µm, but not in a perfectly orderly manner when the tip diameter was 400 µm. These flow regimes corresponded to combined flow regime I, chained flow regime I, separated flow regime I, and chained flow regime II (Figure S1). In addition, under the conditions of high flow rates (3–15 mL/min) and small tip diameters (100–400 µm), two streams aligned in parallel near the tip and deformed chaotically far away from the tip (Movie S2). These results indicate that the flow regime changed from separated flow regime II to turbulence because of increased flow speed. The turbulence is supposed to be caused by two factors: elastic turbulence [23] and inconstant fluidic properties of alginate during cross-linking. We also found that the flow rates required for the transition of flow regimes increased as the tip diameters of the theta-glass capillaries were lengthened. For example, when using theta-glass capillaries with a short tip diameter (100 μm), the transition of the flow regimes from combined flow regime II to separated flow regime II occurred at a low flow rate range (1.0–2.0 mL/min). By contrast, with a middle-length tip diameter (300 μm), the transition occurred at a middle flow rate range (4.0–6.0 mL/min), and with a long tip diameter (500 μm), the transition occurred at a high flow rate range (8.0–10 mL/min). These results indicate that the transition of the flow regimes is a robust phenomenon that occurs regardless of the lengths of the tip diameters of the theta-glass capillaries (i.e., at least within 100–500 μm).

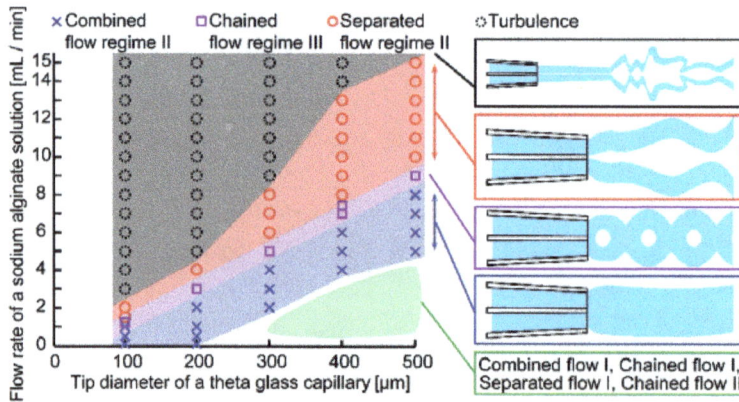

Figure 5. Experimental phases showing relationships between the flow regimes, flow rate of the sodium alginate solution, and tip diameter of the theta-glass capillary. Refer to Figure S1 for detailed data of the bottom-right regime.

3.2. Formation of Branched and Chained Alginate Microfibers

For the results of sodium alginate solution streams in various flow regimes extruded from the theta-glass capillary (500-μm diameter), we obtained alginate microfibers with various shapes, including single-threaded shapes (Figure 6a,b), branched structures (Figure 6c,d), and chained structures (Figure 6e,f). Formed alginate microfibers had different diameters based on the flow regimes. The single-threaded fiber formed in combined flow regime I (0.01 mL/min) (Figure 6a) had a shorter diameter than that of the microfiber formed in combined flow regime II (7.0 mL/min) (Figure 6b). The branched microfiber formed by switching from combined flow regime I (0.01 mL/min) to separated flow regime I (0.1 mL/min) (Figure 6c) also had shorter diameters than that of the microfiber formed by switching from combined flow regime II (7.0 mL/min) to separated flow regime II (11.0 mL/min) (Figure 6d). Similarly, the chained fiber formed in chained flow regime I (0.03 mL/min) (Figure 6e) had a shorter diameter than that of the microfiber formed in chained flow regime II (9.0 mL/min) (Figure 6f). These results indicate that the shapes of the alginate microfibers can be controlled based on the flow rates of the sodium alginate solution with varying diameters. To the best of our knowledge, ours is the first microfluidic method that can be used to form chained alginate microfibers.

To evaluate the variability in the shape of chained alginate microfibers, we formed chained alginate microfibers by extruding 2.0% (*w/w*) sodium alginate solution containing fluorescent beads into a calcium chloride solution through the theta-glass capillary with a 400-μm tip diameter at different flow rates (7.0–8.5 mL/min) in the same flow regime (chained flow regime III) (Figure 7a–d). In this experiment, the range of flow rates generating chained flow regime III (7.0–8.5 mL/min) was different from that shown in Figure 5 because we used the sodium alginate solution with a different concentration (2.0% (*w/w*)) as well as a coloring agent (fluorescent nanobeads). As the flow rate of the sodium alginate solution increased, chain-unit lengths in chained microfibers, defined in Figure 7e, tended to increase. This result indicates that the lengths were variable according to the flow rate because the flow regime continuously changed from a combined flow regime (the length: 0) to a separated flow regime (the length: ∞).

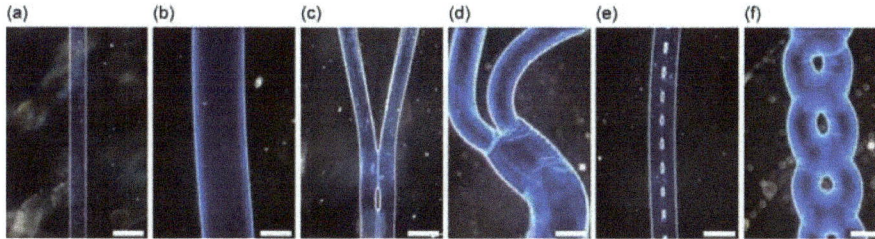

Figure 6. Alginate microfibers with various shapes: (**a**,**b**) single-threaded fibers formed in a combined flow regime (**a**) I and (**b**) II; (**c**,**d**) branched fibers formed in flow regimes changing from (**c**) combined flow regime I to separated flow regime I, and (**d**) combined flow regime II to separated flow regime II; (**e**,**f**) Chained fibers formed in chained flow regime (**e**) I and (**f**) II. Scale bars are 500 μm.

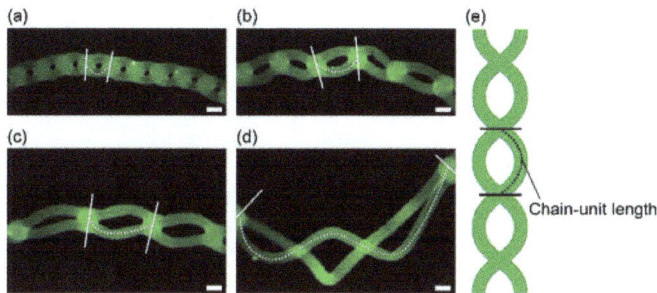

Figure 7. Fluorescent images of chained alginate microfibers with different chain-unit lengths. Microfibers were formed at flow rates of (**a**) 7.0 mL/min; (**b**) 7.5 mL/min; (**c**) 8.0 mL/min; and (**d**) 8.5 mL/min. The dotted line between two lines represents the chain-unit length; (**e**) Conceptual image of chain-unit lengths. Scale bars are 500 μm.

3.3. Microchannel Formation with Chained Alginate Microfibers

To demonstrate the use of chained alginate microfibers as sacrificial templates, we formed a microfiber with a structure where two long and thin fibers are connected to a chained fiber by switching from a separated flow regime to chained flow regime (Figure 8a) and prepared a chained microchannel with two inlets by washing out the microfiber with a sodium citrate solution after embedding the microfiber in a PDMS structure (Figure 8b). The average diameters of the embedded alginate microfiber and the PDMS microchannel in each separated and combined region was 290 ± 18 and 295 ± 18 μm in separated regions, 681 ± 49 and 685 ± 26 μm in combined regions, respectively (Figure S2). The difference between the diameters of the alginate fiber and the PDMS microchannel was within 4% in both separated regions and combined regions. To check the shape of the microchannel, we observed the cross-section of the microchannel by slicing the PDMS structure into thin layers (approximately 200 μm thick). As a result, we confirmed that combined (single-hole channel) and separated (double-hole channel) parts were formed in the microchannel (Figure 8c). These results indicate that the alginate microfiber works as a sacrificial template that transfers its chained structure into the PDMS substrate. Compared to previous hydrogel template methods for forming branched microchannels which requires manual knotting [9] or arranging [8] of two threads of microfibers, our method can form a branched structure utilizing the microfluidic phenomenon, thereby improving the reproducibility of branched and chained shapes.

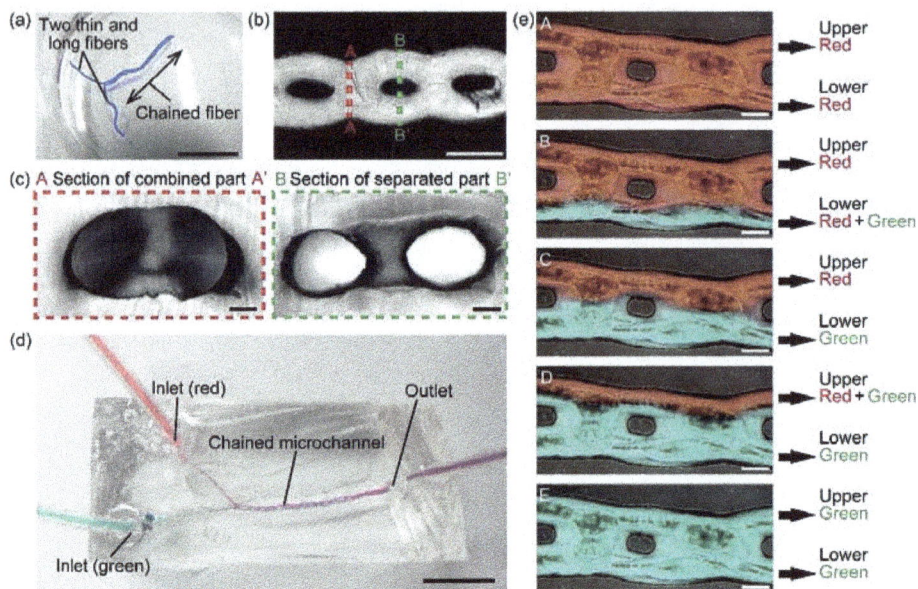

Figure 8. Chained PDMS microchannel fabricated with chained alginate microfibers. (**a**) Alginate microfiber with the structure that two thin and long fibers are connected to a chained fiber as a sacrificial template; (**b**) top view of the chained microchannel; (**c**) cross-sectional image of the microchannel shown in (**b**). Sections AA' and BB' are the combined and separated parts, respectively; (**d**) entire chained microchannel when infusing red and green solutions via different inlets; (**e**) flows of red and green solutions infused at A: 10 and 0, B: 8 and 2, C: 6.7 and 3.3, D: 5 and 5, E: 3.3 and 6.7, F: 2 and 8, and G: 0 and 10 mL/min, respectively. Scale bars are: (**a,d**) 1 cm; (**b**) 500 μm; (**c,e**) 100 μm.

Finally, to confirm the function as a channel, we introduced two solutions colored with red and green ink into the microchannel through two inlets, respectively (Figure 8d). In this experiment, we varied the flow rate ratio of red and green solutions under five conditions: (i) a flow rate ratio of ∞ (10 mL red and 0 mL/min green solutions) (Figure 8eA), (ii) a flow rate ratio of 3 (7.5 mL red and 2.5 mL/min green solutions) (Figure 8eB), (iii) flow rate ratio of 1 (5 mL red and 5 mL/min green solutions) (Figure 8eC), (iv) flow rate ratio of 0.3 (2.5 mL red and 7.5 mL/min green solutions) (Figure 8eD), and (v) flow rate ratio of 0 (0 mL red and 10 mL/min green solutions) (Figure 8eE). As a result, laminar-like flows were formed, leading to five types of flows downstream: (i) both flows in the upper and lower channels were red (Figure 8eA), (ii) one upper flow was red and one lower flow was mixed (Figure 8eB), (iii) one upper flow was red and one lower flow was green (Figure 8eC), (iv) one upper flow was mixed and one lower flow was green (Figure 8eD), and (v) both flows in the upper and lower channels were green (Figure 8eE). The mixture ratio of red and green inks in this chained channel could be continuously variable as a result of controlling the flow rate ratio. These results indicate that a chained microchannel formed with our method is capable of liquid feeding and could be applied to the preparation of mixed solutions with various mixing ratios.

4. Conclusions

In this study, we developed a simple microfluidic method to form alginate microfibers with branched and chained structures using a theta-glass capillary. With our method, we could generate three flow regimes (combined, separated, and chained) by precisely controlling the flow rate of a sodium alginate solution and the tip diameter of the theta-glass capillary. By adjusting the flow regimes, we could change the cross-sections of flows to form alginate microfibers with various shapes such as single-threaded, branched, and chained microfibers. We were also able to change the microfiber dimensions such as the lengths, diameters, and chain-unit lengths in chained microfibers by regulating the flow rate and by changing the tip diameter of the theta-glass capillary. The advantages of our method are as follows: (i) devices can be easily prepared due to the commercial availability of theta-glass capillaries, (ii) our method offers a variety of complicated geometries and dimensions of microfibers, and (iii) branched and chained structures can be reproduced. Moreover, by switching between the different flow regimes, our method could form sacrificial microfibers with controlled positions of branched and re-combined structure; thus, the method has the potential to fabricate microchannels which mimic the structure of blood vessels featured with a multiple-branched structure or a specific chain-like vascular structure (e.g., vascular ring) and are therefore applicable for research fields of fluidics including hemodynamics.

Supplementary Materials: The following is available online at www.mdpi.com/2072-666X/9/6/303/s1: Figure S1: Experimental phases showing the relationships among the flow rates of the sodium alginate solution, the tip diameters of the theta-glass capillaries, and the flow regime in the region with a low flow rate and large tip diameter; Figure S2: Diameters of the embedded microfiber and the PDMS microchannel; Movie S1: Flow regimes of the sodium alginate solution; Movie S2: Turbulent stream of the sodium alginate solution.

Author Contributions: K.N., Y.M., and N.M. conceived and designed the experiments; K.N. performed the experiments and analyzed the data and wrote the original draft; S.T supervised the overall coordination; All authors discussed the results and contributed to the manuscript.

Acknowledgments: This research is partially supported by Grant-in-Aid for Scientific Research (S) (Grant number: 16H06329) and Grant-in-Aid for JSPS Fellows (Grant number: 18J23131), JSPS KAKENHI.

Conflicts of Interest: The authors declare no conflicts of interest.

References

1. Lee, K.Y.; Mooney, D.J. Alginate: Properties and biomedical applications. *Prog. Polym. Sci.* **2012**, *37*, 106–126. [CrossRef] [PubMed]
2. Morimoto, Y.; Onuki, M.; Takeuchi, S. Mass Production of Cell-Laden Calcium Alginate Particles with Centrifugal Force. *Adv. Healthc. Mater.* **2017**, *6*, 1–5. [CrossRef] [PubMed]
3. Elliott, R.B.; Escobar, L.; Tan, P.L.J.; Muzina, M.; Zwain, S.; Buchanan, C. Live encapsulated porcine islets from a type 1 diabetic patient 9.5 yr after xenotransplantation. *Xenotransplantation* **2007**, *14*, 157–161. [CrossRef] [PubMed]
4. Andersen, T.; Auk-Emblem, P.; Dornish, M. 3D Cell Culture in Alginate Hydrogels. *Microarrays* **2015**, *4*, 133–161. [CrossRef] [PubMed]
5. Silva, J.M.; Custódio, C.A.; Reis, R.L.; Mano, J.F. Multilayered Hollow Tubes as Blood Vessel Substitutes. *ACS Biomater. Sci. Eng.* **2016**, *2*, 2304–2314. [CrossRef]
6. Venkatesan, J.; Bhatnagar, I.; Manivasagan, P.; Kang, K.H.; Kim, S.K. Alginate composites for bone tissue engineering: A review. *Int. J. Biol. Macromol.* **2015**, *72*, 269–281. [CrossRef] [PubMed]
7. Liu, Y.; Sakai, S.; Taya, M. Engineering tissues with a perfusable vessel-like network using endothelialized alginate hydrogel fiber and spheroid-enclosing microcapsules. *Heliyon* **2016**, *2*. [CrossRef] [PubMed]
8. Hirama, H.; Odera, T.; Torii, T.; Moriguchi, H. A lithography-free procedure for fabricating three-dimensional microchannels using hydrogel molds. *Biomed. Microdevices* **2012**, *14*, 689–697. [CrossRef] [PubMed]
9. Mori, N.; Morimoto, Y.; Takeuchi, S. Vessel-like channels supported by poly-L-lysine tubes. *J. Biosci. Bioeng.* **2016**, *122*, 753–757. [CrossRef] [PubMed]
10. Onoe, H.; Takeuchi, S. Cell-laden microfibers for bottom-up tissue engineering. *Drug Discov. Today* **2015**, *20*, 236–246. [CrossRef] [PubMed]

11. Jun, Y.; Kang, E.; Chae, S.; Lee, S.H. Microfluidic spinning of micro- and nano-scale fibers for tissue engineering. *Lab Chip* **2014**, *14*, 2145–2160. [CrossRef] [PubMed]
12. Cheng, J.; Jun, Y.; Qin, J.; Lee, S.H. Electrospinning versus microfluidic spinning of functional fibers for biomedical applications. *Biomaterials* **2017**, *114*, 121–143. [CrossRef] [PubMed]
13. Bhattarai, N.; Li, Z.; Edmondson, D.; Zhang, M. Alginate-based nanofibrous scaffolds: Structural, mechanical, and biological properties. *Adv. Mater.* **2006**, *18*, 1463–1467. [CrossRef]
14. Saquing, C.D.; Tang, C.; Monian, B.; Bonino, C.A.; Manasco, J.L.; Alsberg, E.; Khan, S.A. Alginate-polyethylene oxide blend nanofibers and the role of the carrier polymer in electrospinning. *Ind. Eng. Chem. Res.* **2013**, *52*, 8692–8704. [CrossRef]
15. Tottori, S.; Takeuchi, S. Formation of liquid rope coils in a coaxial microfluidic device. *RSC Adv.* **2015**, *5*, 33691–33695. [CrossRef]
16. Nie, M.; Takeuchi, S. Microfluidics based synthesis of coiled hydrogel microfibers with flexible shape and dimension control. *Sens. Actuators B Chem.* **2017**, *246*, 358–362. [CrossRef]
17. Cheng, Z.; Cui, M.; Shi, Y.; Qin, Y.; Zhao, X. Fabrication of cell-laden hydrogel fibers with controllable diameters. *Micromachines* **2017**, *8*, 161. [CrossRef]
18. Onoe, H.; Okitsu, T.; Itou, A.; Kato-Negishi, M.; Gojo, R.; Kiriya, D.; Sato, K.; Miura, S.; Iwanaga, S.; Kuribayashi-Shigetomi, K.; et al. Metre-long cell-laden microfibres exhibit tissue morphologies and functions. *Nat. Mater.* **2013**, *12*, 584–590. [CrossRef] [PubMed]
19. Hsiao, A.Y.; Okitsu, T.; Onoe, H.; Kiyosawa, M.; Teramae, H.; Iwanaga, S.; Kazama, T.; Matsumoto, T.; Takeuchi, S. Smooth muscle-like tissue constructs with circumferentially oriented cells formed by the cell fiber technology. *PLoS ONE* **2015**, *10*, 1–16. [CrossRef] [PubMed]
20. Kang, E.; Jeong, G.S.; Choi, Y.Y.; Lee, K.H.; Khademhosseini, A.; Lee, S.H. Digitally tunable physicochemical coding of material composition and topography in continuous microfibres. *Nat. Mater.* **2011**, *10*, 877–883. [CrossRef] [PubMed]
21. Kang, E.; Choi, Y.Y.; Chae, S.K.; Moon, J.H.; Chang, J.Y.; Lee, S.H. Microfluidic spinning of flat alginate fibers with grooves for cell-aligning scaffolds. *Adv. Mater.* **2012**, *24*, 4271–4277. [CrossRef] [PubMed]
22. Ji, X.; Guo, S.; Zeng, C.; Wang, C.; Zhang, L. Continuous generation of alginate microfibers with spindle-knots by using a simple microfluidic device. *RSC Adv.* **2015**, *5*, 2517–2522. [CrossRef]
23. Grolsman, A.; Stelnberg, V. Elastic turbulence in a polymer solution flow. *Nature* **2000**, *405*, 53–55. [CrossRef] [PubMed]

micromachines

MDPI

Article

Development of a Triple-Coaxial Flow Device for Fabricating a Hydrogel Microtube and Its Application to Bioremediation

Kazuma Fujimoto [1], Kazuhiko Higashi [1] , Hiroaki Onoe [2] and Norihisa Miki [2,*

[1] School of Integrated Design Engineering, Keio University, 3-14-1 Hiyoshi, Kohoku-ku, Yokohama, Kanagawa 223-8522, Japan; fkazuma2007@gmail.com (K.F.); kazuhiko@z2.keio.jp (K.H.)

[2] Department of Mechanical Engineering, Keio University, Yokohama 223-8522, Japan; onoe@mech.keio.ac.jp (H.O.)

* Correspondence: miki@mech.keio.ac.jp; Tel.: +81-45-566-1430

Received: 10 January 2018; Accepted: 9 February 2018; Published: 12 February 2018

Abstract: This paper demonstrates a triple-coaxial flow device to continuously produce a hydrogel microtube using a microfluidic technique. The hydrogel microtube can encapsulate a microbial suspension, while allowing the diffusion of oxygen and nutrients into the microtube and preventing microbes from passing into or out of the microtube. The microtubes also enable the collection of the microbes after task completion without contaminating the environment. In our previous study, we used a double-coaxial flow device to produce the microtubes, but continuous production was a challenge. In the present study, we developed a microfluidic device that fabricates a triple-coaxial flow to enable continuous production of the microtubes. Here, we characterize the production capacity of the microtubes along with their properties and demonstrate bioremediation using microtubes encapsulating a microbial suspension.

Keywords: microtubes; triple-coaxial flow; microbes; microfluidics; bioremediation

1. Introduction

Microbes conduct multistep reactions with biological enzymes as catalysts for metabolization [1]. Many of the resulting substances of the metabolism are difficult to produce in vitro; therefore, microbes are currently used in various fields such as food production, medicine, environmental science, and energy [2–7]. For practical applications using microbes, the culture system needs to enable mass culture at low cost without biological contamination from competitive microbes.

In our previous work, we proposed a microbial culture system with hydrogel microtubes, as shown in Figure 1 [8]. The hydrogel tubes are made of calcium alginate, and have pores that are larger than nutrients and oxygen, but smaller than microbes and bacteriophages [9,10]. Therefore, the hydrogel microtubes prevent microbes from passing through the walls while permitting the diffusion of oxygen and nutrients. Target microbes encapsulated inside the tubes are, thus, protected from competing microbes. Moreover, the microbes can be easily collected along with the tubes without contaminating the surrounding environment. We previously developed a microfluidic device to produce the microtubes that encapsulate microbes using double-coaxial flow [11–13]. The double-coaxial flow comprises a sodium alginate solution as the outer flow and a microbial suspension as the inner flow. When sodium alginate bonds with calcium ions, it immediately forms a stable three-dimensional gel [14,15]. As shown in Figure 2, the double-coaxial flow flows into the calcium chloride solution and the outer flow, sodium alginate, becomes a hydrogel to form the hydrogel microtube encapsulation. The flow rates and, thus, the production rate, are limited due to

Rayleigh–Taylor instability and continuous formation of the tubes is difficult [16,17]. As previously mentioned, mass production is mandatory for practical applications.

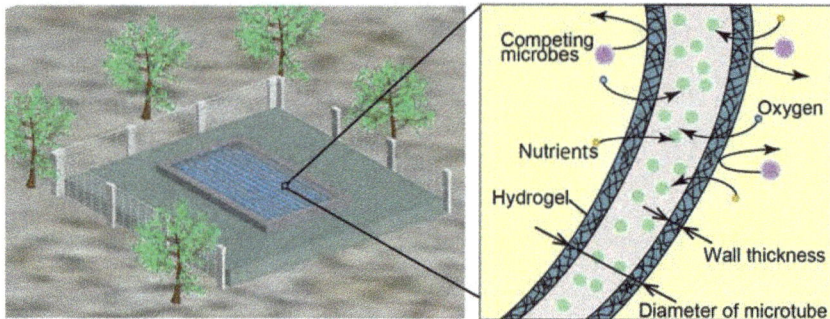

Figure 1. Open culture system with microtubes.

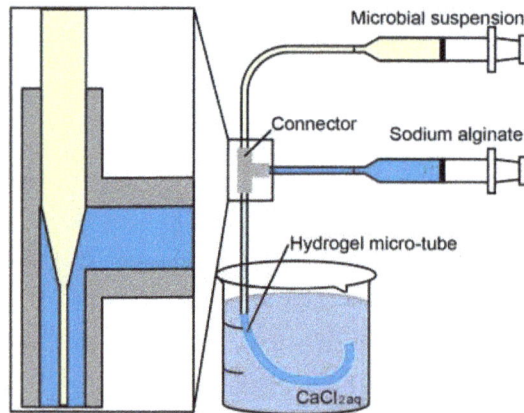

Figure 2. Our previous device with the double-coaxial flow.

In this study, therefore, we developed a triple-coaxial flow device, where the outermost flow is calcium chloride, and hydrogel microtubes emerge from the end of the device. The device enables mass production of the microtubes with negligible Rayleigh–Taylor instability. We first characterize the formation of the hydrogel tubes with respect to the production rate, efficiency, and the tube properties. Then, we demonstrate bioremediation, which is one of the promising microbial applications, using the microtubes with encapsulated microbes. The microtubes were immersed in an aqueous solution of methylene blue to decompose the methylene blue, and when the bioremediation was completed, the tubes and the microbes inside were successfully collected.

2. Materials and Methods

2.1. Materials

Sodium alginate (80–120 cP) and calcium chloride were purchased from Wako Pure Chemical Industries, Ltd. (Osaka, Japan), and saline was obtained from Otsuka Pharmaceutical Factory (Tokyo, Japan). The concentration of sodium alginate solution was 1.5 w/v% (7.5 g sodium alginate in 492.5 g deionized (DI) water). The concentration of the calcium chloride solution was 150 mM (16.65 g calcium chloride in 1000 mL DI water). These conditions were found to be appropriate for producing hydrogel microtubes [9,18]. *Coryne glutamicum* (*C. glutamicum*), *Vibrio alginolyticus* (*V. alginolyticus*), *Pseudomonas aeruginosa* (*P. aeruginosa*), and *Bacillus subtilis* (*B. subtilis*) were purchased from the National Institute of Technology and Evaluation (Tokyo, Japan) and were used as the target microbes. The culture media of *C. glutamicum*, *P. aeruginosa*, and *B. subtilis* included 5 g hipolypepton, 1 g yeast extract, 0.5 g $MgSO_4 \cdot 7H_2O$, and 500 mL DI water. The culture medium of *V. alginolyticus* included 5 g hipolypepton, 1 g yeast extract, 0.25 g $MgSO_4 \cdot 7H_2O$, 13.35 g Daigo's artificial seawater SP (NOVA Chemicals, Calgary, Canada), and 500 mL DI water. All the materials for producing the different culture media were purchased from Wako Pure Chemical Industries. Methylene blue aqueous solution was used to demonstrate bioremediation [19,20], and was prepared by mixing 0.029 g methylene blue trihydrate powder (Hayashi Pure Chemical Ind., Ltd., Osaka Japan) and 500 mL DI water. The solution for the experiments of the removal of methylene blue dye included 15 mL methylene blue solution, 50 mL DI water, and 20 mL culture medium

2.2. Triple-Coaxial Flow Device

A photo and the cross-sectional view of the triple-coaxial flow device are shown in Figure 3. The device consists of three aluminum parts made using a lathe and numerically-controlled milling machine (Figure 3a). As shown in Figure 3b,c, the triple-coaxial flow consists of three fluids: (#1) the microbial suspension, (#2) the sodium alginate solution, and (#3) the calcium chloride solution. A hydrogel microtube encapsulating the microbial suspension is produced by the formation of the triple-coaxial flow from the three fluids, and emerges from the outlet of the device. To observe the effect of the outlet diameter, we designed separate Part 2 sections of the device with three different outlet diameters.

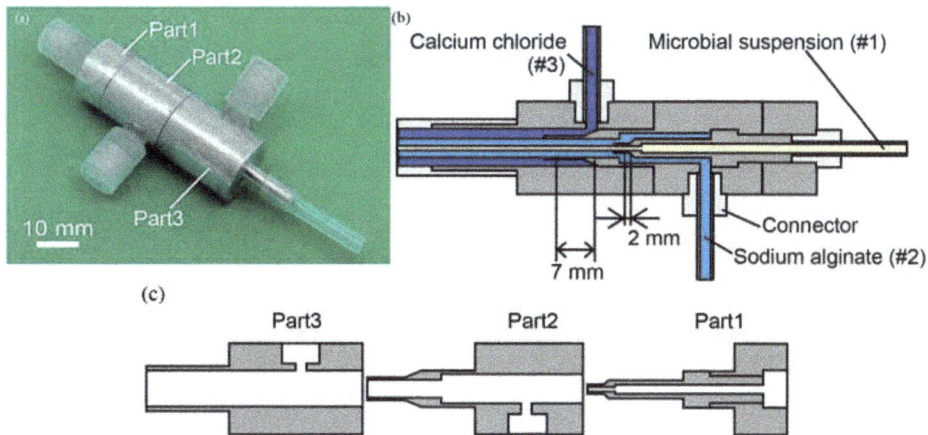

Figure 3. Triple-coaxial flow device. (**a**) Image of the device. Cross-sectional view of (**b**) the device and (**c**) the three parts to produce the triple-coaxial flow.

The experimental setup is shown in Figure 4. The triple-coaxial flow device was fixed to a stand. The microbial suspension was introduced into the device using a liquid delivery pump (Peri-star pump Pro, Tacmina Co., Osaka, Japan). The sodium alginate solution was introduced into the device using a pulseless pump (Smoothflow pump Q, Tacmina Co., Osaka, Japan). The calcium chloride solution was introduced into the device using a liquid delivery pump (Masterflex L/S, Yamato Scientific Co, Ltd., Tokyo, Japan). Each pump and the device were connected using silicone tubing. Pulsatile flow prevents a stable formation of the triple-coaxial flow, so an air chamber was placed between the pumps and the device to reduce the pulsations generated by the liquid delivery pumps. When we stop producing the microtube, #1, #2, and #3 flows are stopped in turn.

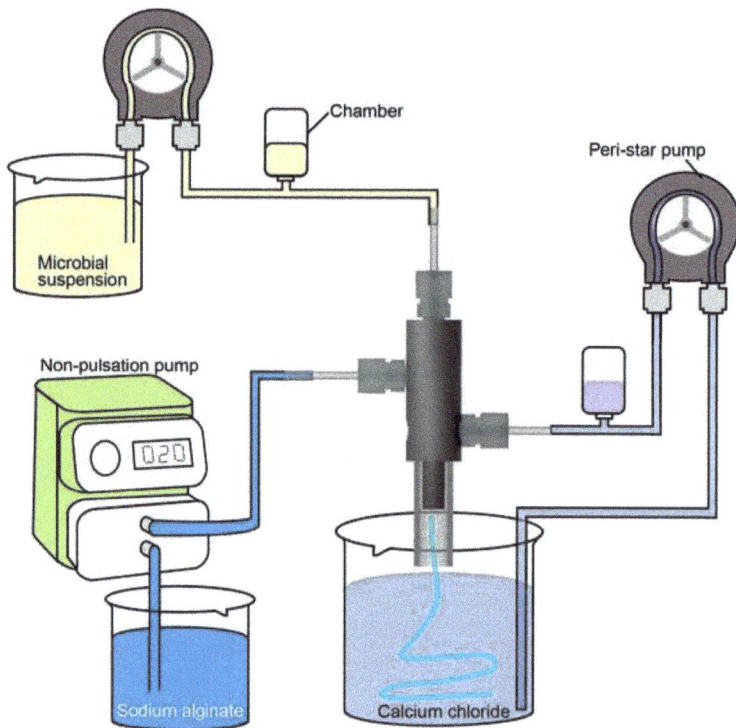

Figure 4. Experimental setup for production of microtubes.

2.3. Fabrication of a Hollow Hydrogel Microtube

Saline dyed with red stain was used as fluid #1 instead of a microbial suspension to characterize the formation of the microtubes. We investigated the wall thickness of the microtube with respect to the volumetric flow rates of fluids #1 and #2. The volumetric flow rate of fluid #1 was set from 12.5 to 27.5 mL/min, the volumetric flow rate of fluid #2 was set from 25 to 35 mL/min, and the volumetric flow rate of fluid #3 was kept constant at 75 mL/min. Fluids #3, #2, and #1 were sequentially introduced into the device. The microtubes were subsequently collected in the calcium chloride solution. The outer and inner diameters of the fabricated microtubes were measured at three locations using an optical microscope (VHX-600, KEYENCE, Osaka, Japan).

2.4. Bioremediation: Removal of Methylene Blue Dye

First, we conducted preliminary experiments to determine the suitable microbes for bioremediation. We used methylene blue to visualize the ability of the microbes to decompose organic matter. The tested microbes included *C. glutamicum*, *V. alginolyticus*, *P. aeruginosa*, and *B. subtilis* [21–23]. To investigate the capability of these microbes to decompose methylene blue, 15 mL of each microbial suspension was added to a solution consisting of 15 mL methylene blue solution and 70 mL DI water. To evaluate the degree of methylene blue decomposition, we measured changes in the absorbance of methylene blue with a spectrophotometer (UV3600, Shimadzu, Kyoto, Japan) after one day, four days, seven days, and 14 days. The absorbance was measured three times for each solution.

Next, we produced a hydrogel microtubes encapsulating 15 mL of the microbial suspension of *B. subtilis* following the results of the preliminary tests. The produced microtube was immersed into the methylene blue solution. After one day, four days, and seven days, the absorbance and the degree of the turbidity of the methylene blue aqueous solution was measured three times using the spectrophotometer. We also attempted to collect *B. subtilis* along with the microtubes to confirm that collection of encapsulated microbes is possible without contaminating the solution with the microbes.

3. Results and Discussion

3.1. Fabrication of a Hollow Hydrogel Microtube

Figure 5 shows the film thickness of the produced microtube with respect to the volumetric flow rate. As the volumetric flow rate of fluid #1 increased, the inner diameter of the microtube increased, but the film thickness decreased. It was also found that the film thickness increased as the volumetric flow rate of fluid #2 increased. Thereupon, it was found that the film thickness was controlled from 100 to 500 μm. We investigated the effect of the outlet diameter of Part 2. Figure 6 shows the outer diameter of the microtube produced by the triple-coaxial flow device with respect to the volumetric flow rate of fluid #2. The microtube diameter increased as both the outer diameter of Part 2 and the volumetric flow rate of fluid #2 increased. Furthermore, it was found that the diameter of the hydrogel microtube was controlled from 1.4 to 2.0 mm. In our previous study we found that, to ensure diffusion of nutrients and oxygen through the tube wall, the wall thickness needed to be less than 250 μm. Moreover, for mass culture of target microbes, the diameter of the microtube should be as large as possible. Therefore, we chose volumetric flow rates of 27.5, 35, and 75 mL/min for fluids #1, #2, and #3, respectively. This is because of the produced microtube at the volumetric flow rates. A photo of the produced hydrogel microtube encapsulating dyed saline is shown in Figure 7.

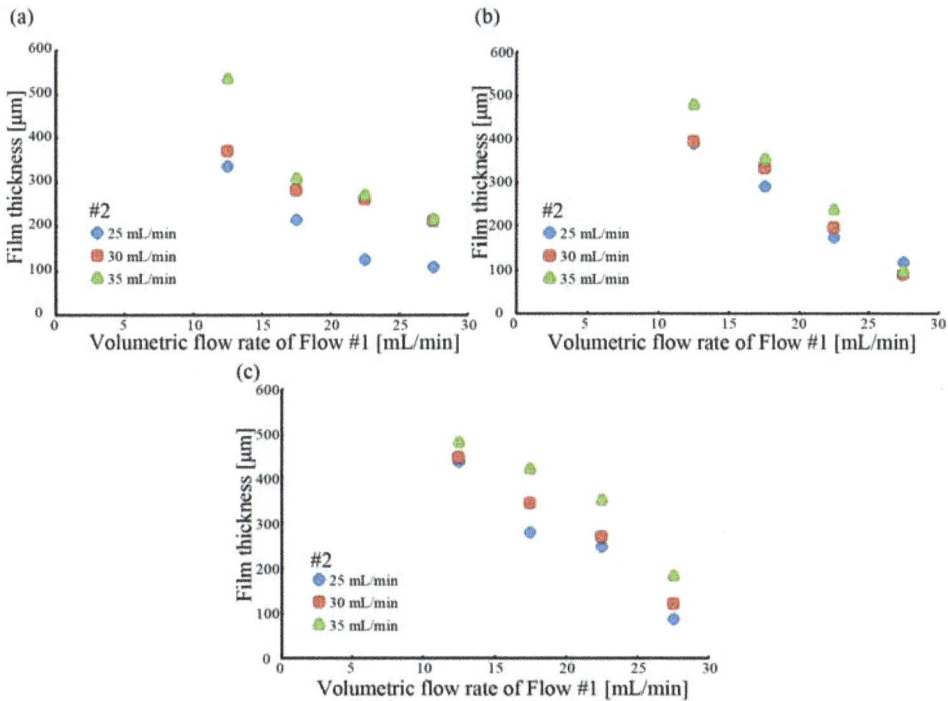

Figure 5. Relationship between film thickness of microtube and volumetric flow rates of fluids #1 and #2. (**a**) Outlet diameter of Part 2 is 2.5 mm. (**b**) Outlet diameter of Part 2 is 2.7 mm. (**c**) Outlet diameter of Part 2 is 2.9 mm.

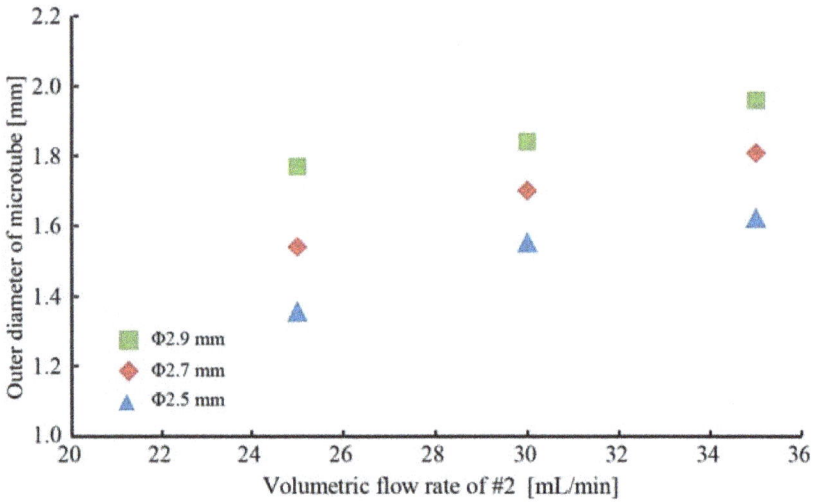

Figure 6. Relationship between outer diameter of microtube and volumetric flow rate of fluid #2.

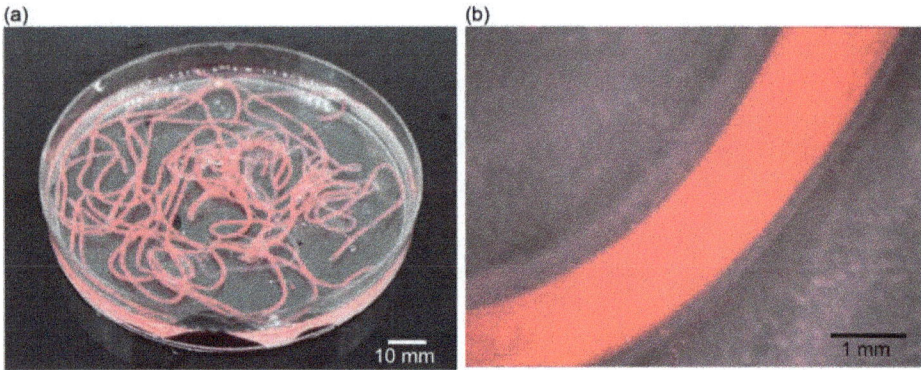

Figure 7. (a) Photo of the produced hydrogel microtube with the triple-coaxial device. **(b)** Photomacrogragh of the hydrogel microtube.

Microtubes were produced at a rate of 281 mm/s. This is approximately 25 times faster than the 11 mm/s achieved in our previous study using a double-coaxial flow device. Next, we measured the outlets at 10 locations on microtubes produced using double- and triple-coaxial flow devices. Figure 8 shows the average diameter and the standard error of the microtubes produced using the two devices. As we described in the Introduction, the Rayleigh–Taylor instability generated during fabrication of the microtubes using the double-coaxial flow device results in the formation of a hydrogel mass. Therefore, the standard error of the outlet diameter of the microtubes was large because of the formation of this hydrogel mass. The triple-coaxial flow can produce the hydrogel tubes with sufficient mechanical strength before they reach the calcium chloride solution, or collecting solution. These differences resulted in higher production rates and smaller deviations in the tube diameter.

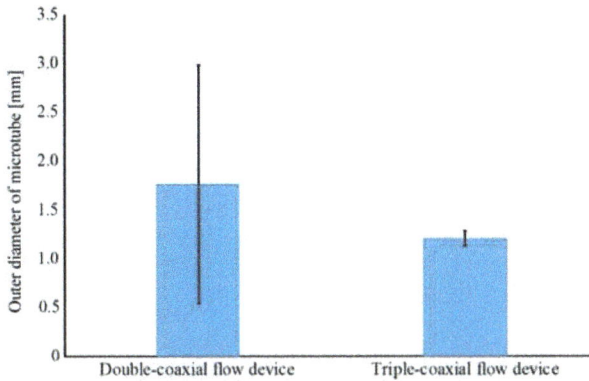

Figure 8. Comparison of outer diameters of microtubes devices produced using the double- and triple-coaxial flow.

3.2. Removal of Methylene Blue Dye

Figure 9 shows the degree of decomposition of methylene blue for each of the target microbes. *V. alginolyticus*, *P. aeruginosa*, and *B. subtilis* decomposed the methylene blue, while *C. glutamicum* did not. Among the tested microbes, *B. subtilis* exhibited the highest performance in decomposition of methylene blue. Based on the results of this experiment, *B. subtilis* could have reduced the concentration of methylene blue in the aqueous solution by 46% in two weeks. Therefore, we decided to use *B. subtilis* as the target microbe in further experiments testing the effectiveness of encapsulated microbes. Notably, the standard error of the absorbance ratio for each microbe increased as the culture time progressed. This is because the turbidity caused by the increase in the number of microbes interferes with the absorbance measurements of methylene blue.

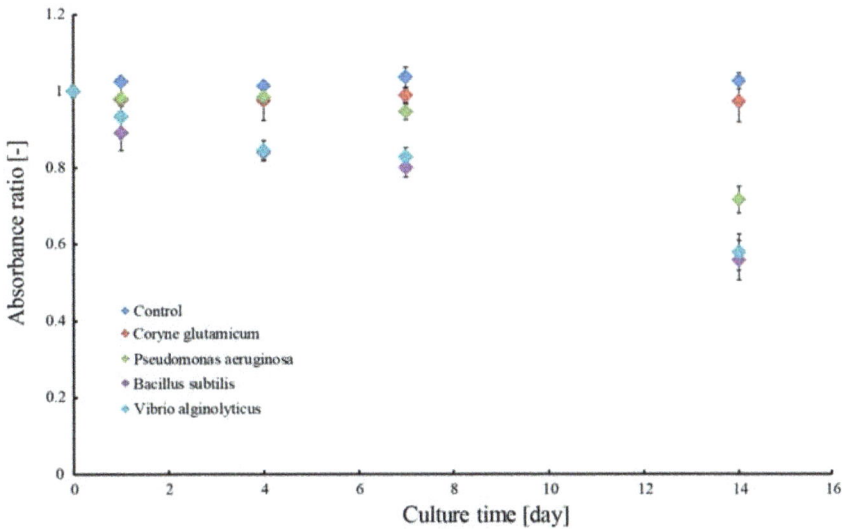

Figure 9. Relationship between absorbance ratio of methylene blue and culture time of each microbe.

Next, we produced microtubes encapsulating a microbial suspension of *B. subtilis*. The produced microtubes were immersed into the methylene blue aqueous solution. Figure 10 shows the degree of decomposition of methylene blue using the hydrogel microtubes. The hydrogel microtubes encapsulating the microbial suspension decomposed methylene blue at a comparable rate to the microbes that were not encapsulated. It is considered that *B. subtilis* encapsulated in the hydrogel microtube can obtain enough oxygen and nutrients.

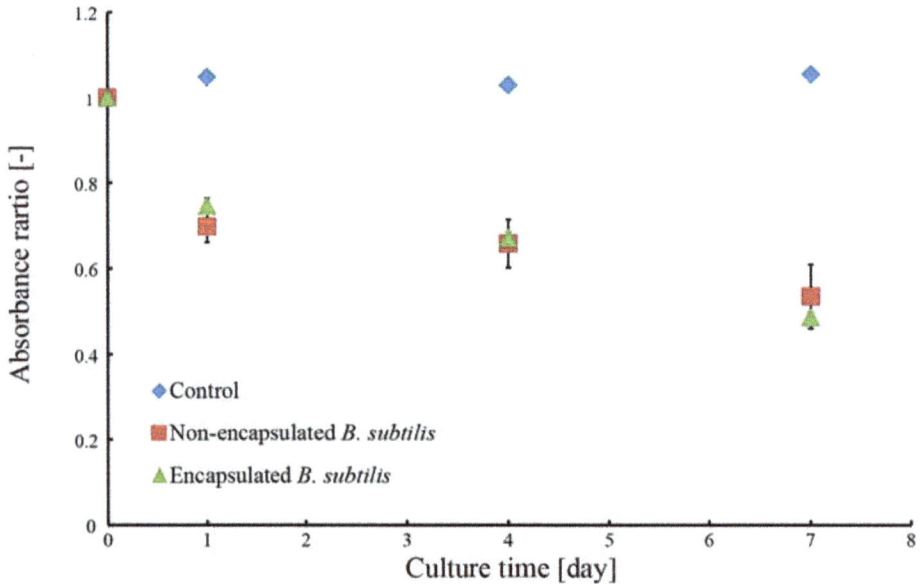

Figure 10. Relationship between the absorbance ratio of methylene blue and culture time with encapsulated and non-encapsulated *B. subtilis*.

Figure 11 shows the opacity of the methylene blue aqueous solution during the bioremediation experiments by measuring the changes in the absorbance ratio of the methylene blue aqueous solution. When *B. subtilis* was introduced into the solution, the opacity of the solution increased due to the presence of the microbes. However, when the microtubes containing the microbes were introduced into the solution, the value of the absorbance ratio of the concentration of microbes had fallen in the range 0.95–1.05. The absorbance ratio of the control solution, which did not contain microbes, had fallen in the same range. Therefore, the opacity did not increase. This indicates that the microbes did not leak out of the microtubes. Again, the higher standard error of the absorbance ratio was caused by the turbidity of the solution containing *B. subtilis* not encapsulated in the microtubes. Furthermore, the microtubes were successfully collected and removed from the solution along with the microtubes, which prevented the contamination of the solution by the microbes. This can only be achieved by encapsulating the microbes in the microtubes. As shown in Figure 10, the encapsulated microbes could decompose the methylene blue. *B. subtilis* encapsulated in the hydrogel microtubes reduced the concentration of the methylene blue solution by 50% in seven days.

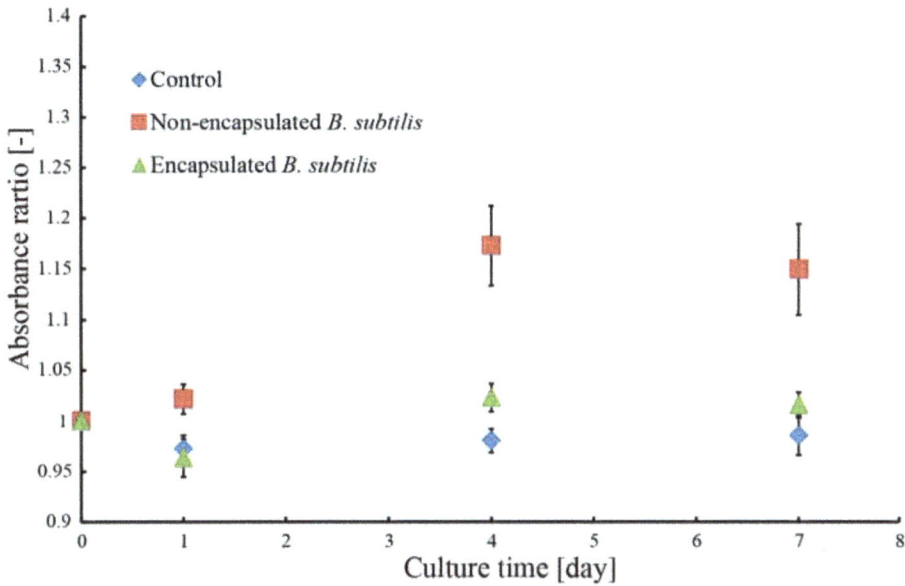

Figure 11. Relationship between the concentration of microbes in the methylene blue solution and culture time.

4. Conclusions

We developed and characterized a triple-coaxial flow device to produce hydrogel microtubes continuously at a high rate. The diameter and the film thickness of the hydrogel microtubes were successfully controlled with the volumetric flow rates of the fluids and the diameter of the channel. Therefore, by using the triple-coaxial flow device, the diameter of the hydrogel microtube was controlled from 1.4 mm to 2.0 mm, and the film thickness was controlled from 100 μm to 500 μm. The triple-coaxial flow device was able to produce microtubes 25 times faster than our previously-proposed double-coaxial flow device. By reducing the effect of Rayleigh–Taylor instability, the produced hydrogel microtubes had smoother features and less deviation in their diameter. The effectiveness of the hydrogel microtubes encapsulating microbes was demonstrated in bioremediation experiments. *B. subtilis* encapsulated inside the microtubes successfully decomposed methylene blue and did not leak out of the tubes, thereby preventing contamination of the solution by the microbes. Moreover, the microbes were easily collected from the solution along with the microtubes.

Author Contributions: Kazuma Fujimoto and Hiroaki Onoe designed and manufactured the triple-coaxial device. Kazuma Fujimoto, Kazuhiko Higashi, and Norihisa Miki designed the experiments. Kazuma Fujimoto analyzed the data. Kazuma Fujimoto and Norihisa Miki wrote the paper.

Conflicts of Interest: The authors declare no conflict of interest.

References

1. Ogimoto, K. *Basic Knowledge of Microorganisms for Biotechnology: Microorganisms Surrounding Humans*; Kodansha: Tokyo, Japan, 2012; ISBN 9784061537309.
2. Ogawa, T.; Tamori, M.; Kimura, A.; Mine, A.; Sakuyama, H.; Yoshida, E.; Maruta, T.; Suzuki, K.; Ishikawa, T.; Shigeoka, S. Enhancement of photosynthetic capacity in Euglena gracilis by expression of cyanobacterial fructose-1,6-/sedoheptulose-1,7-bisphosphatase leads to increases in biomass and wax ester production. *Biotechnol. Biofuels* **2015**, *8*. [CrossRef] [PubMed]

3. Shimada, R.; Fujita, M.; Yuasa, M.; Sawamura, H.; Watanabe, T.; Nakashima, A.; Suzuki, K. Oral administration of green algae, Euglena gracilis, inhibits hyperglycemia in OLETF rats, a model of spontaneous type 2 diabetes. *Food Funct.* **2016**, *7*, 4655–4659. [CrossRef] [PubMed]
4. Borowitzka, M.A. Commercial production of microalgae: Ponds, tanks, tubes and fermenters. *J. Biotechnol.* **1999**, *70*, 313–321. [CrossRef]
5. Spolaore, P.; Joannis-Cassan, C.; Duran, E.; Isambert, A. Commercial applications of microalgae. *J. Biosci. Bioeng.* **2006**, *101*, 87–96. [CrossRef] [PubMed]
6. Fahnestock, S.R.; Yao, Z.; Bedzyk, L.A. Microbial production of spider silk proteins. *J. Biotechnol.* **2000**, *74*, 105–119. [CrossRef]
7. Heslot, H. Artificial fibrous proteins: A review. *Biochimie* **1998**, *80*, 19–31. [CrossRef]
8. Ogawa, M.; Higashi, K.; Miki, N. Microbial production inside microfabricated hydrogel microtubes. In Proceedings of the 18th International Conference on Solid-State Sensors, Actuators and Microsystems (TRANSDUCERS), Anchorage, AK, USA, 21–25 June 2015; pp. 1692–1694.
9. Hirayama, K.; Okitsu, T.; Teramae, H.; Kiriya, D.; Onoe, H.; Takeuchi, S. Cellular building unit integrated with microstrand-shaped bacterial cellulose. *Biomaterials* **2013**, *34*, 2421–2427. [CrossRef] [PubMed]
10. Boggione, D.M.G.; Batalha, L.S.; Gontijo, M.T.P.; Lopez, M.E.S.; Teixeira, A.V.N.C.; Santos, I.J.B.; Mendonca, R.C.S. Evaluation of microencapsulation of the UFV-AREG1 bacteriophage in alginate-Ca microcapsules using microfluidic devices. *Colloids Surf. B Biointerfaces* **2017**, *158*, 182–189. [CrossRef] [PubMed]
11. Higashi, K.; Ogawa, M.; Fujimoto, K.; Onoe, H.; Miki, N. Hollow hydrogel microfiber encapsulating microorganisms for mass-cultivation in open systems. *Micromachines* **2017**, *8*, 176. [CrossRef]
12. Ogawa, M.; Higashi, K.; Miki, N. Development of hydrogel microtubes for microbe culture in open environment. In Proceedings of the 2015 37th Annual International Conference of the IEEE Engineering in Medicine and Biology Society, New York, NY, USA, 25–29 August 2015; pp. 5896–5899.
13. Fujimoto, K.; Higashi, K.; Onoe, H.; Miki, N. Microfluidic mass production system for hydrogel microtubes for microbial culture. *Jpn. J. Appl. Phys.* **2017**, *56*. [CrossRef]
14. Onoe, H.; Takeuchi, S. Cell-laden microfibers for bottom-up tissue engineering. *Drug Discov. Today* **2015**, *20*, 236–246. [CrossRef] [PubMed]
15. Osada, Y.; Kajihara, K. *Gel Handbook*; NTS Inc.: Tokyo, Japan, 1997; p. 64. ISBN 4860430298.
16. He, X.; Zhang, R.; Chen, S.; Doolen, G.D. On the three-dimensional Rayleigh–Taylor instability. *Phys. Fluids Phys. Fluids A Fluid Dyn.* **1999**, *11*, 1143–1152. [CrossRef]
17. Wysocki, A.; Royall, C.P.; Winkler, R.G.; Gompper, G.; Tanaka, H.; van Blaaderen, A.; Löwen, H. Direct observation of hydrodynamic instabilities in a driven non-uniform colloidal dispersion. *Soft Matter* **2009**, *5*, 1340–1344. [CrossRef]
18. Onoe, H.; Okitsu, T.; Itou, A.; Kato-Negishi, M.; Gojo, R.; Kiriya, D.; Sato, K.; Miura, S.; Iwanaga, S.; Kuribayashi-Shigetomi, K.; et al. Metre-long cell-laden microfibres exhibit tissue morphologies and functions. *Nat. Mater.* **2013**, *12*, 584–590. [CrossRef] [PubMed]
19. Li, F.; Li, X.; Zhang, J.D.; Peng, L.; Liu, C.Y. Removal of organic matter and heavy metals of low concentration from wastewater via micellar-enhanced ultrafiltration: An overview. *IOP Conf. Ser. Earth Environ. Sci.* **2017**, *52*. [CrossRef]
20. Ai, L.; Jiang, J. Removal of methylene blue from aqueous solution with self-assembled cylindrical graphene-carbon nanotube hybrid. *Chem. Eng. J.* **2012**, *192*, 156–163. [CrossRef]
21. Pacwa-Płociniczak, M.; Płaza, G.A.; Piotrowska-Seget, Z.; Cameotra, S.S. Environmental applications of biosurfactants: Recent advances. *Int. J. Mol. Sci.* **2011**, *12*, 633–654. [CrossRef] [PubMed]
22. Wang, J.; Li, Y.; Bian, C.; Tong, J.; Fang, Y.; Xia, S. Ultramicroelectrode array modified with magnetically labeled Bacillus subtilis, palladium nanoparticles and reduced carboxy graphene for amperometric determination of biochemical oxygen demand. *Microchim. Acta* **2017**, *184*, 763–771. [CrossRef]
23. Upendar, G.; Dutta, S.; Chakraborty, J.; Bhattacharyya, P. Removal of methylene blue dye using immobilized bacillus subtilis in batch & column reactor. *Mater. Today Proc.* **2016**, *3*, 3467–3472. [CrossRef]

micromachines

MDPI

Article

Crack-Configuration Analysis of Metal Conductive Track Embedded in Stretchable Elastomer

Tomoya Koshi [ID] and Eiji Iwase *[ID]

Department of Applied Mechanics, Waseda University, 3-4-1 Okubo, Shinjuku-ku, Tokyo 169-8555, Japan; koshi@akane.waseda.jp
* Correspondence: iwase@waseda.jp; Tel.: +81-03-5286-2741

Received: 28 February 2018; Accepted: 9 March 2018; Published: 15 March 2018

Abstract: This paper reports the analysis of the crack configuration of a stretched metal conductive track that is embedded in a stretchable elastomer. The factor determining the crack configurations is analyzed by modeling as well as experiments. The modeling analysis indicates that the crack configuration is determined by the ratio of the elongation stiffness of the track and elastomer, and is classified into two types: multiple-crack growth and single-crack growth. When the track stiffness is considerably lower than the elastomer stiffness, multiple-crack growth type occurs; in the opposite case, single-crack growth type occurs. Hence, to verify the modeling analysis, metal conductive tracks with different thicknesses are fabricated, and the cracks are studied with respect to the crack width, number of cracks, and crack propagation speed. In this study, two conventional metal-track shapes are studied: straight-shaped tracks with track thickness of 0.04–1.17 µm, and wave-shaped tracks with track thickness of 2–10 µm. For straight-shaped tracks, multiple-crack growth type occurred, when the track thickness was 0.04 µm, and the crack configuration gradually changed to a single crack, with the increase in the track thickness. For wave-shaped tracks with track thickness of 2–10 µm, only single-crack growth type occurred; however, the crack propagation speed decreased and the maximum stretchability of the track increased, with the increase in the track thickness.

Keywords: crack configuration; metal conductive track; stretchable elastomer; flexible electronic device; stretchable electronic device

---

1. Introduction

Of late, many research groups have been developing flexible or stretchable electronic devices [1–3], such as stretchable displays [4,5], devices fixed to the human skin [6–11], and neural interfaces devices that are embedded in animals [12]. As metal conductive tracks are one of the critical components for achieving device flexibility or stretchability, various types of metal tracks, such as straight-shaped metal tracks with microcracks [13–16], straight-shaped metal tracks on a wavy surface [17,18], and wave-shaped metal tracks [19–22], have been researched. Straight-shaped tracks with microcracks are stretchable and conductive with randomly distributed tribranched microcracks on the tracks; the track thickness is several tens or hundreds of nanometers, and the metal track layer is directly deposited on a stretchable elastomer substrate by thermal or electron-beam deposition [12]. Straight-shaped metal tracks on a wavy surface are fabricated by direct metal deposition on a prestretched elastomer substrate and can be stretched by the deformation of the wavy surface of the elastomer [17]. Wave-shaped metal tracks can be stretched by the deformation of the wave-shape. The thickness of the metal layer is serval micrometers, and it is fabricated by plating or laminating a metal foil and a stretchable elastomer sheet [19,22].

In previous studies, the observed crack configurations of stretched metal tracks differ considerably. For example, many micro cracks were observed in some studies [12–16]; whereas, few large cracks, which propagated and crossed the metal track perpendicular to the stretching direction, were observed in the others [19–21]. However, the factors determining the crack configurations are still not clear. Understanding these factors can contribute to the development and improvement of flexible or stretchable electronic devices because crack configurations affect both the flexibility and stretchability of a metal conductive track. In addition, this understanding can contribute to related studies, such as self-healing methods for a cracked metal track [23,24].

In this study, to analyze the factors that determine the crack configurations, modeling as well as experiments were utilized. In the experiments, we studied two conventional metal-track shapes: straight-shaped metal tracks and wave-shaped metal tracks. For straight-shaped tracks, thinner tracks with thickness of several tens to hundreds of nanometers are generally used to achieve the stretchability using out-of-plane deformation by tribranched microcracks or a wavy surface. For wave-shaped tracks, the in-plane deformation of the wave-shape is mainly used for stretchability; hence, thicker tracks with a thickness of several micrometers are generally used. Therefore, in this study, straight-shaped tracks with track thickness of 0.04–1.17 µm and wave-shaped tracks with track thickness of 2–10 µm were fabricated, and the cracks were studied with respect to the crack width, number of cracks, and crack propagation speed.

2. Material and Methods

2.1. Modeling Analysis

The factors determining the crack configurations were analyzed by modeling. Figure 1a displays the schematic of a cracked metal track embedded in a stretchable elastomer. A small crack is caused in the track, and both the track and the elastomer are deformed by a constant balanced force. In the model, the strain in the cracked region is denoted as ε_A, and that in the non-cracked region as ε_B. We assumed that all of the strains are uniform over each region, and that ε_A is larger than ε_B, as shown in Figure 1b,c. These figures depict the simplified stress-strain curves of a metal track and stretchable elastomer, respectively. For flexible or stretchable electronic devices, gold or copper are used as conductive tracks, and polydimethylsiloxane (PDMS) or polyurethane (PU) are used as stretchable elastomer layers. Therefore, the stress-strain curves are simplified, based on the material. In the model, the balanced force around the boundary between the cracked and non-cracked regions is represented by

$$\varepsilon_A E_{elast} A_{elast} = \sigma_{break} A_{track} + \varepsilon_B E_{elast} A_{elast} \tag{1}$$

where E_{elast}, A_{elast}, A_{track}, and σ_{break} are the Young's modulus of the elastomer, cross-sectional area of the elastomer, cross-sectional area of the track, and breaking stress of the track, respectively. Focusing on ε_A and ε_B, Equation (1) is represented as

$$\varepsilon_A - \varepsilon_B = \frac{\sigma_{break} A_{track}}{E_{elast} A_{elast}} \tag{2}$$

The left of Equation (2) is the difference between ε_A and ε_B, and the right is the ratio of the elongation stiffness of the track and elastomer. Therefore, Equation (2) indicates that the difference between ε_A and ε_B is determined by the ratio of the elongation stiffness. In particular, when the ratio of the elongation stiffness is considerably small, then the stiffness of the track is considerably lower than that of the elastomer, i.e.,

$$\varepsilon_A - \varepsilon_B \approx 0 \tag{3}$$

In this case, the values of ε_A and ε_B are nearly the same; hence, we consider that other cracks are caused in the non-cracked region, as the track is stretched further, as shown in Figure 1d. In this paper, we refer to this crack configuration as a multiple-crack growth type. On the other hand, when the ratio of the elongation is greater than zero, we consider that a crack already caused in the track increasingly propagates, as the track is stretched further, as shown in Figure 1e. We refer to this crack configuration as a single-crack growth type.

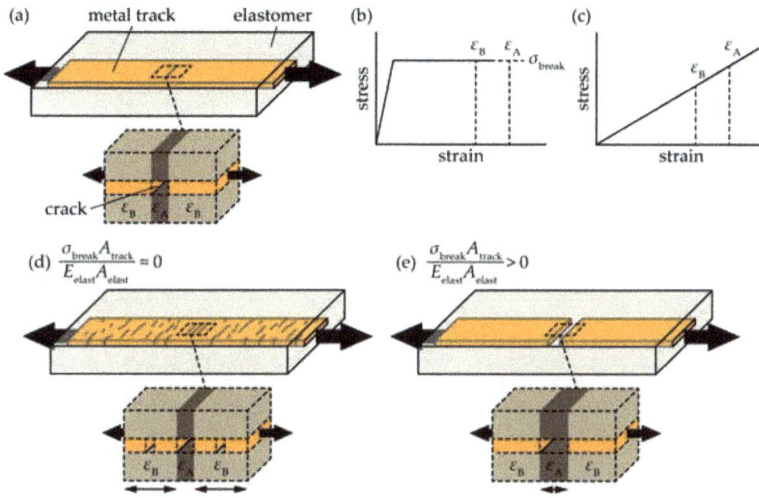

Figure 1. Schematic of the crack-configuration modeling analysis. (**a**) Cracked metal track embedded in a stretchable elastomer; (**b**) simplified stress-strain curve of a metal track layer; (**c**) simplified stress strain-curve of a stretchable elastomer; (**d**) schematic of a multiple-crack growth type; and, (**e**) schematic of a single-crack growth type.

To examine the relationship between the crack configurations and the ratio of the elongation stiffness, the value of the ratio was calculated, with reference to previous studies. In the calculation, we assumed that the widths of the track and elastomer are considerably greater than the thicknesses of the track and elastomer; hence, Equation (2) is represented as

$$\varepsilon_A - \varepsilon_B = \frac{\sigma_{break} t_{track}}{E_{elast} t_{elast}} \tag{4}$$

where t_{track} and t_{elast} are the thicknesses of the track and elastomer, respectively. Table 1 shows the calculated values of the ratio of the elongation stiffness, using Equation (4); the values that are used in Table 1 are as follows: σ_{break} of gold was 0.3 GPa [25], σ_{break} of copper was 0.2 GPa [26], and E_{elast} of PDMS was 1.3 MPa [27]. Table 1 indicates that the multiple-crack growth type is often observed, when the value of the ratio of the elongation stiffness is lower than approximately 0.1. On the other hand, the single-crack growth type is often observed, when the value of the ratio is higher than approximately unity.

Table 1. Relationship between the calculated values of the ratio of the elongation stiffness and the crack configurations, in previous studies and this study.

Ref.	Structure	$\frac{\sigma_{break} t_{track}}{E_{elast} t_{elast}}$	Crack Configuration
[13] (straight-shaped track)	t_{track} = 0.05 µm (gold) t_{elast} = 1 mm (PDMS)	0.01	
[16] (straight-shaped track)	t_{track} = 0.05–0.1 µm (gold) t_{elast} = 1 mm (PDMS)	0.01–0.02	
[14] (straight-shaped track)	t_{track} = 0.075 µm (gold) t_{elast} = 0.3 mm (PDMS)	0.06	Multiple-crack growth type
[12] (straight-shaped track)	t_{track} = 0.035 µm (gold) t_{elast} = 0.12 mm (PDMS)	0.07	
[15] (straight-shaped track)	t_{track} = 0.04 µm (gold) t_{elast} = 0.076 mm (PDMS)	0.12	

Table 1. *Cont.*

Ref.	Structure	$\frac{\sigma_{break}\,t_{track}}{E_{elast}\,t_{elast}}$	Crack Configuration
[19] (straight-shaped/wave-shaped track)	t_{track} = 2.5–5 μm (gold) t_{elast} = 0.4 mm (PDMS)	1.44–2.88	Single-crack growth type
[21] (wave-shaped track)	t_{track} = 18 μm (copper) t_{elast} = 1 mm (PDMS)	2.76	
[20] (wave-shaped track)	t_{track} = 17 μm (copper) t_{elast} = 0.1 mm (PDMS)	26.15	
This study (straight-shaped track)	t_{track} = 0.04–1.17 μm (copper) t_{elast} = 0.1 mm (PU)	0.03–0.78	Multiple-crack growth/Single-crack growth type
This study (wave-shaped track)	t_{track} = 2–10 μm (copper) t_{elast} = 0.1 mm (PU)	0.89–4.45	Single-crack growth type

2.2. Fabrication and Experimental Setup

To verify the modeling analysis, metal conductive tracks with different ratios of the elongation stiffness were fabricated. This ratio was varied by changing the thickness of the metal track. In this study, two conventional metal-track shapes were studied: straight-shaped tracks with track thicknesses of 0.04–1.17 μm, and wave-shaped tracks with track thickness of 2–10 μm. In previous studies on straight-shaped tracks, a metal track layer was deposited on a stretchable elastomer layer by thermal or electron-beam deposition. However, a stretchable elastomer, such as PDMS and PU, changes its mechanical property around the boundary between the track and elastomer, due to thermal damage. In this case, the mechanical property is unclear, and it is difficult to comprehend the factors determining the crack configurations. In both the thermal deposition and transfer methods, some microcracks might be pre-formed in the metal layer; however, the transfer methods is much better because of the no thermal damage. Moreover, if there are the pre-formed microcracks, then the pre-formed microcracks have little effect on the crack configuration. For the multiple-crack growth type, new microcracks are caused in addition to the pre-formed microcracks. On the other hand, for the single-crack growth type, some of the pre-formed microcracks propagate and become the large cracks. Therefore, in this study, a transfer method, in which the metal layer was transferred onto the elastomer, was used. For the wave-shaped metal track, commercially available rolled copper foil was used as the metal track layer. PU was used as the elastomer layer, for both straight-shaped and wave-shaped tracks.

Figure 2a–e depict the fabrication process of a straight-shaped track. Initially, a polytetrafluoroethylene (PTFE) sheet was cut into 20 mm × 30 mm sheets, and a copper layer was deposited on these PTFE sheets by a thermal evaporation system (SVC-700TMSG/7PS80, Sanyu Electron Co., Ltd., Tokyo, Japan), as shown in Figure 2a. The track thickness, t_{track}, was 0.04 μm, 0.10 μm, 0.18 μm, 0.53 μm, and 1.17 μm, respectively. Further, the PTFE sheet was cut in the shape of a track with a width of 3 mm (Figure 2b), and was pasted onto a PU tape (Tegaderm, 3M, Meipplewood, MN, USA). The thickness of the PU tape was 0.05 mm. The PTFE sheet was then peeled off from the PU tape, and the copper layer was transferred from the PTFE sheet to the PU tape (Figure 2c). Finally, another PU tape was pasted on the copper layer, as shown in Figure 2d, and the PU layer was cut into a sample shape with a width of 10 mm (Figure 2e). The calculated ratios of the elongation stiffness of each sample were 0.03, 0.07, 0.12, 0.35, and 0.78, at t_{track} = 0.04 μm, 0.10 μm, 0.18 μm, 0.53 μm, and 1.17 μm, respectively. In the calculation, an E_{elast} of 3 MPa was used for the PU tape.

Figure 3a–g show the fabrication process of a wave-shaped copper track. Initially, rolled copper foils (The Nilaco Co., Tokyo, Japan) of various thicknesses (t_{track} = 2 μm, 4 μm, 6 μm, 8 μm, and 10 μm) were thermally laminated onto a PU sheet (Platilon 4201, Covestro AG, Leverkusen, Germany), as shown in Figure 3a. The thickness of the PU sheet was 0.05 mm. In the lamination process, the copper foil was surface-modified by a plasma cleaner (PDC-32G, Harrick Plasma, New York, NY, USA); the copper foil and PU sheet were heated at 100 °C for 3 min without pressure, and was then pressed at approximately 0.4 MPa, at 170 °C for 3 min. Subsequently, the copper foil was structured by the photolithography process. Photoresist was spin-coated onto the copper foil (Figure 3b), and patterned into a wave-shaped track and contact pads (Figure 3c). Further, the copper layer was wet-etched (Figure 3d) and the photoresist was stripped off (Figure 3e). A PU tape was laminated only on the wave-shaped track, as shown in Figure 3f. Finally, the individual samples were separated. Figure 3e shows the fabricated wave-shaped copper track. The dimensions of each sample were 25 mm by 5 mm. The width of the copper track was 75 μm, and it was arranged between two large contact pads, which were 5 mm apart. The radius of the track was 150 μm. The calculated ratio of the elongation stiffness of each sample was 0.89, 1.78, 2.67, 3.56, and 4.45 at t_{track} = 2 μm, 4 μm, 6 μm, 8 μm, and 10 μm, respectively. In the calculation, 3 MPa and 6 MPa were used as the E_{elast} values of the PU tape and PU sheet, respectively.

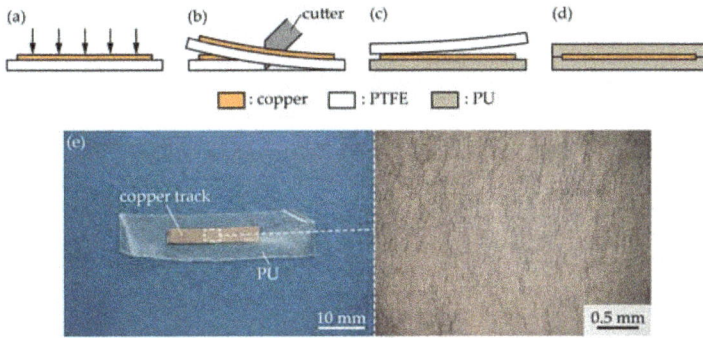

Figure 2. Fabrication of a straight-shaped copper track embedded in PU. (**a**) Thermal deposition of a copper layer on a polytetrafluoroethylene (PTFE) sheet; (**b**) Cutting of the PTFE sheet; (**c**) Transfer of the copper track onto a polyurethane (PU) tape; (**d**) Lamination of another PU tape; and, (**e**) Optical images of the fabricated copper track in PU.

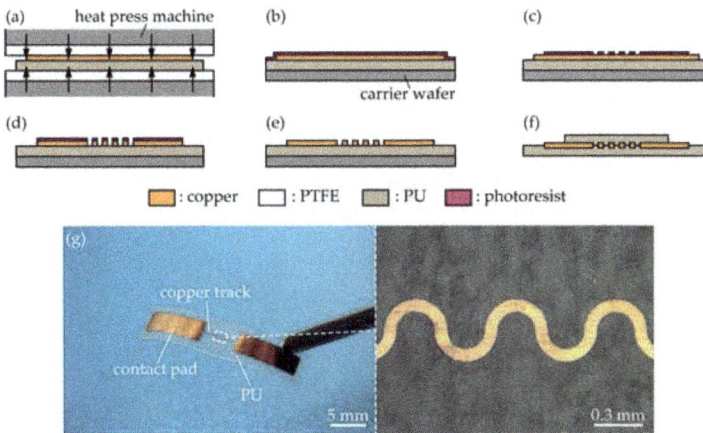

Figure 3. Fabrication of a wave-shaped copper track embedded in PU. (**a**) Lamination of a copper foil on a PU sheet; (**b**) Spin-coating of a photoresist on the copper foil; (**c**) Development and patterning of the photoresist; (**d**) Wet-etching of the copper foil; (**e**) Removal of the photoresist; (**f**) Lamination of a PU tape on the structured copper; and, (**g**) Optical images of a wave-shaped copper track in PU.

Figure 4 displays images of the experimental setup. A sample were mounted onto movable stages. For wave-shaped tracks, the sample was clamped at the contact pads, and electronically connected to a source meter (2614B, Keithley Instruments, Cleveland, OH, USA). The resistance of the wave-shaped track was measured by four probe methods, for detecting crack propagation. The sample was stretched gradually by moving the stages. The crack formation was observed with an optical microscope (VHX-2000, Keyence Corporation, Osaka, Japan).

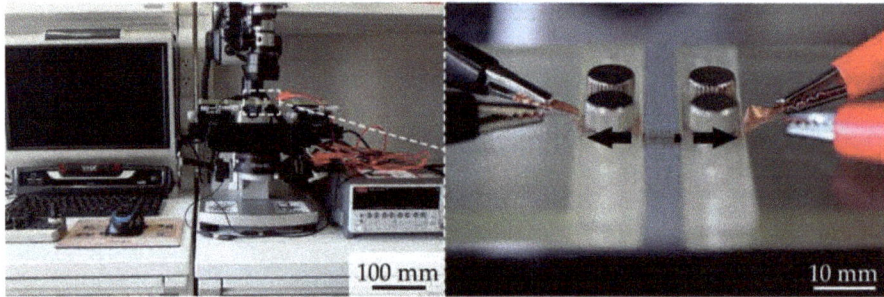

Figure 4. Images of the experimental setup.

3. Result and Discussion

3.1. Straight-Shaped Track

Figure 5a–i depict a series of optical images of a cracked track. For a stretched track with t_{track} = 0.04 µm (the calculated ratio of the elongation stiffness was 0.03), several smaller cracks were observed in the track (Figure 5a–c). When the elongation rate was 10%, few cracks were caused in the track, as shown in Figure 5a. As the elongation rate increased, new cracks were caused, and the number of cracks increased (Figure 5b,c). This indicates that a crack configuration with t_{track} = 0.04 µm is a multiple-crack, corresponding to the results of Table 1. In this case, the track might be conductive even if the track was stretched up to several tens percent as shown in a previous study [15]. For t_{track} = 0.53 µm (0.35 is the ratio of the elongation stiffness), several larger cracks were observed in the track, as shown in Figure 5g–i. The cracks propagated along the track-width direction, which was nearly perpendicular to the stretching direction. The crack width increased with the increase in the elongation rate; whereas, the number of cracks was nearly constant. For t_{track} = 1.17 µm (0.78 is the ratio of the elongation stiffness), a similar trend was observed. This indicates that the crack configuration for t_{track} = 0.53 µm and 1.17 µm is single-crack growth type, and also corresponds to the results of Table 1. In this case, the track might lose its conductivity, even when the elongation rate was under several percent, because of the larger crack propagation. For a cracked track with t_{track} = 0.10 µm and 0.18 µm, an intermediate type of crack configuration was observed, as shown in Figure 5d–f. Many microcracks were caused, and each crack propagated, as the elongation rate increased.

For better understanding, numerical analysis on the crack width and number of cracks was conducted. The crack width and number of cracks were measured from the optical image of a cracked track (Figure 6a,b), and the transition of the crack width and the number of cracks were analyzed. The crack width was measured as follows: seven cracks were randomly selected on the optical image of a cracked track, and the values of each crack area were measured. The crack width was obtained by dividing each area by the height of each crack. Hence, the crack width is the average value, along the stretching direction. The number of the cracks was calculated as follows: seven lines was drawn on an optical image of a cracked track at regular intervals, and the number of cracks, across the line, were counted. The direction of the line was along the stretching direction, and the line was drawn end-to-end on the optical image. The counted number was divided by the reference distance, which was 100 μm for an elongation of 0%. Therefore, the number of cracks is the average value of the reference distance. The crack widths for t_{track} = 0.04 μm, 0.10 μm, and 0.18 μm ranged from several micro to several tens of micrometers, as shown in Figure 6c. On the other hand, for t_{track} = 0.53 μm and 1.17 μm, the crack widths ranged from several tens to hundreds of micrometers. Figure 6d shows the normalized crack width by the value of the crack width at 10% elongation rate. For t_{track} = 0.04 μm, the normalized crack width ranged from 1 to 2. On the other hand, for t_{track} = 0.10 μm, 0.18 μm, 0.53 μm, and 1.17 μm, the normalized crack width increased almost linearly, as the elongation rate increased. This indicates that in the case of the multiple-crack growth type, the crack width is nearly constant or increases marginally, as the elongation rate increases; whereas, in the case of a single-crack growth type, the crack width increases almost linearly. The number of cracks per reference distance was more than one, for t_{track} = 0.04 μm, 0.10 μm, and 0.18 μm (Figure 6e). On the other hand, for t_{track} = 0.53 μm and 1.17 μm, the number of cracks were approximately zero. Figure 6f shows the normalized number of cracks by the value of the number of cracks at a 10% elongation rate. For t_{track} = 0.04 μm, the normalized number of cracks increased from approximately 1–90, as the elongation rate increased. On the other hand, for t_{track} = 0.10 μm, 0.18 μm, 0.53 μm, and 1.17 μm, the normalized number was nearly constant, at unity. That indicates that, in the case of the multiple crack-growth type, the number of cracks increases suddenly, as the elongation rate increases; whereas, in the case of the single-crack growth type, the number is nearly constant.

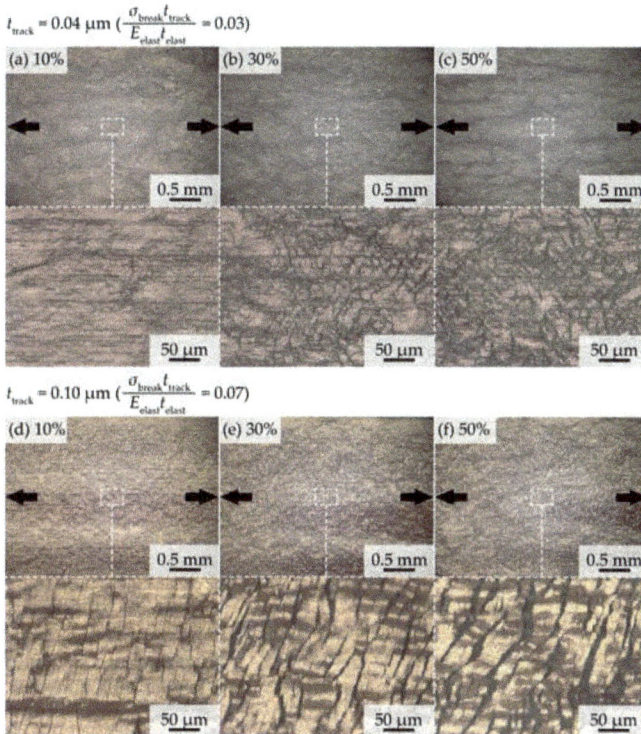

Figure 5. *Cont.*

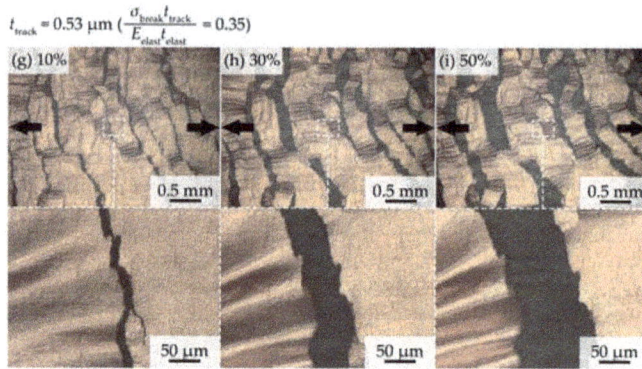

Figure 5. Series of optical images of a cracked copper track in PU with track thicknesses of (**a**–**c**) 0.04 μm; (**d**–**f**) 0.53 μm; and, (**g**–**i**) 0.10 μm.

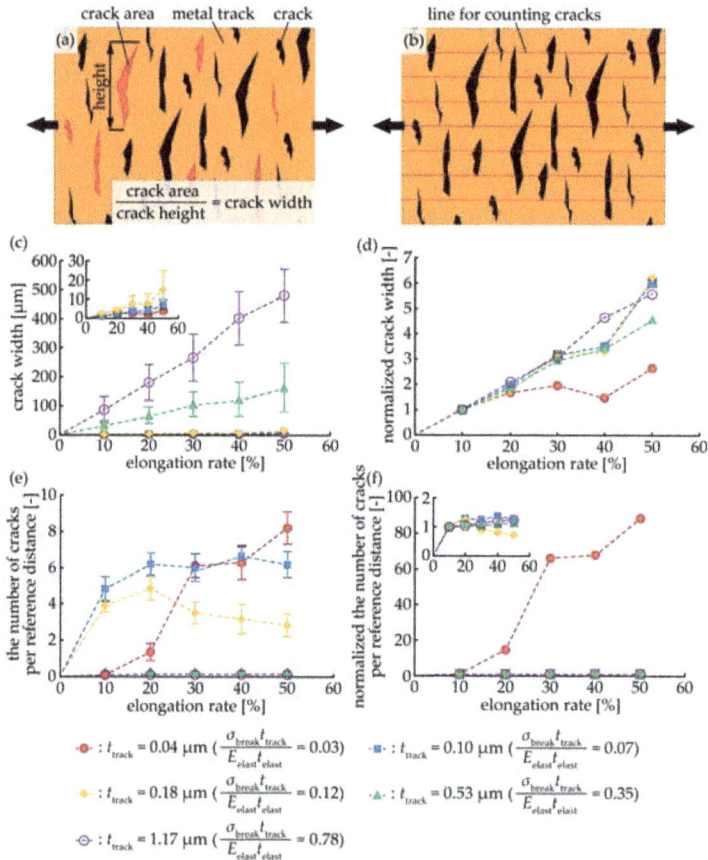

Figure 6. Schematic of the (**a**) measurement of the crack width and (**b**) number of cracks. Relationship between the (**c**) crack width and elongation rate; (**d**) normalized crack width and elongation rate; (**e**) number of cracks and elongation rate; and, (**f**) normalized number of cracks and elongation rate. The reference distance was 100 μm for an elongation of 0%.

3.2. Wave-Shaped Track

Figure 7a shows the resistance-change rate in terms of the elongation rate, for t_{track} = 2 μm and 6 μm (0.89 and 1.78 are the calculated ratios of the elongation stiffness, respectively). In both cases, the resistance rate increased, as the elongation rate increased, and each track completely broke at 13% and 42% of the elongation rate for t_{track} = 2 μm and 6 μm, respectively. Figure 7b–g show the optical images of the crack propagation of each track. When strain was gradually applied to the track, some crack initiations were observed (Figure 7b,e), around the apex of the wave shape. Each crack gradually propagated as the elongation rate increased (Figure 7c,f), along the direction of the track width. This indicates that the resistance rate was increased by crack propagation. The tracks were completely broken, when the crack propagated completely (Figure 7d,g). The crack configuration was the single-crack growth type, in both cases, corresponding to the results of Table 1. When compared with t_{track} = 2 μm and 6 μm, the crack propagation speed was different. Figure 7a–g indicate that the crack propagation speed at t_{track} = 2 μm was faster than that at 6 μm.

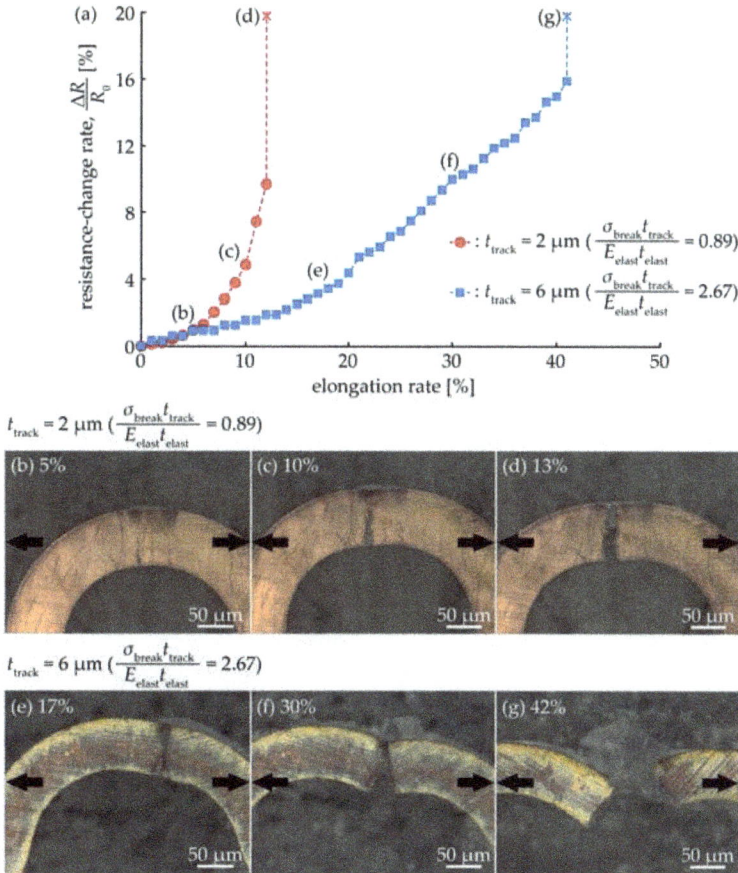

Figure 7. (a) Relationship between the resistance-change rate and elongation rate for t_{track} = 2 μm and 4 μm; Optical images around the apex of a wave shape for (**b**–**d**) t_{track} = 2 μm and (**e**–**g**) t_{track} = 6 μm.

Figure 8a shows the relationship between the maximum stretchability of a wave-shaped track and the track thickness, t_{track}. The number of trials for each thickness was five. The values of the maximum stretchability were 12%, 20%, 35%, 40%, and 50% at 2 μm, 4 μm, 6 μm, 8 μm, and 10 μm of t_{track}, respectively; therefore, the maximum stretchability increased proportionally, as t_{track} increased. This indicates that the crack propagation speed was reduced, as t_{track} increased; hence, the maximum stretchability increased. In addition to this, we consider that the maximum stretchability may also be increased by decreasing the width or wavelength of the wave-shape as shown in previous studies [19,22]. Figure 8b shows the relationship between the crack width and t_{track}. The number of trials for each thickness was five, again. The crack was trapezoid-shaped; therefore, the crack width was defined as the distance between the mid points of the sides. The values of the crack width were 16 μm, 36 μm, 66 μm, 69 μm, and 94 μm at 2 μm, 4 μm, 6 μm, 8 μm, and 10 μm of t_{track}, respectively; hence, the crack width increased proportionally, as the track thickness increased. This indicates that a larger strain energy was caused, as the track elongation rate increased, and this energy was released when the track was broken. This larger energy renders the crack wider.

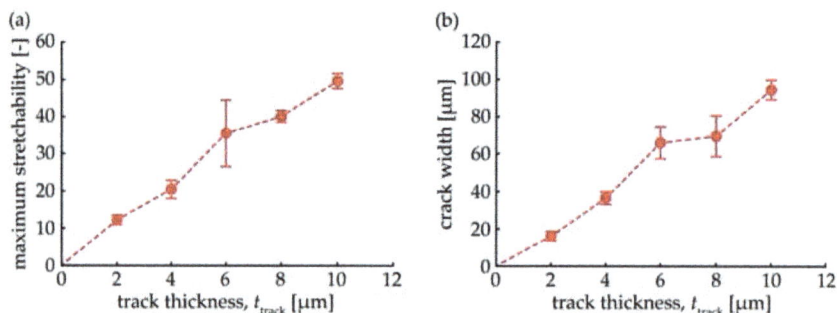

Figure 8. Relationship between the (**a**) maximum stretchability and track thickness; t_{track}, and the (**b**) crack width and track thickness, t_{track}.

4. Conclusions

We analyzed the factors determining the crack configurations of a stretched metal conductive track embedded in stretchable elastomer, by both modeling and experiments. The modeling analysis indicated that the crack configuration is determined by the ratio of the elongation stiffness of the track and the elastomer, and it is classified into two types: multiple-crack growth and single-crack growth. We established that the multiple-crack growth type is often observed, when the ratio of the elongation stiffness is lesser than approximately 0.1, and a single-crack growth type is often observed, when ratio is more than approximately unity, with reference to previous studies. In the experiments, to verify the modeling analysis, two conventional metal-track shapes were examined: straight-shaped tracks with t_{track} = 0.04–1.17 μm (0.03–0.78 is the ratio of the elongation stiffness), and wave-shaped tracks with t_{track} = 2–10 μm (0.89–4.45 is the ratio of the elongation stiffness). For straight-shaped tracks, the multiple-crack growth type was observed, when t_{track} = 0.04 μm, and the crack configuration gradually changed to a single crack, as t_{track} increased. This corresponds to the result of the modeling analysis. For wave-shaped tracks, only a single-crack growth type was observed; hence, this also corresponds to the modeling analysis. In addition, for a wave-shaped track, the crack propagation speed was reduced and the maximum stretchability of the track increased linearly, with the increase in t_{track}.

Acknowledgments: This work was partially supported by JST CREST, Grant Number: JPMJCR16Q5 and JSPS KAKENHI, Grant Number: 16KT0107. The authors would like to acknowledge Thomas Löher and his colleagues at Fraunhofer IZM for their support and valuable discussions, in this research. The authors thank the MEXT Nanotechnology Platform Support Project of Waseda University.

Author Contributions: T.K. and E.I. conceived and designed the experiments; T.K. performed the experiments, analyzed the data, and wrote the paper; E.I. supervised the research.

Conflicts of Interest: The authors declare no conflict of interest.

References

1. Hammock, M.L.; Chortos, A.; Tee, B.C.K.; Tok, J.B.H.; Bao, Z. 25th anniversary article: The evolution of electronic skin (E-skin): A Brief history, design considerations, and recent progress. *Adv. Mater.* **2013**, *25*, 5997–6038. [CrossRef] [PubMed]
2. Kim, D.H.; Xiao, J.; Song, J.; Huang, Y.; Rogers, J.A. Stretchable, curvilinear electronics based on inorganic materials. *Adv. Mater.* **2010**, *22*, 2108–2124. [CrossRef] [PubMed]
3. Rogers, J.A.; Someya, T.; Huang, Y. Materials and mechanics for stretchable electronics. *Science* **2010**, *327*, 1603–1607. [CrossRef] [PubMed]
4. Ohmea, H.; Tomita, Y.; Kashara, M.; Schram, J.; Smits, E.; Brand, J.; Bossuyt, F.; Vanfleteren, J.; Baets, J.D. Stretchable 45 × 80 RGB LED display using meander wiring technology. *SID Symp. Dig. Tech. Pap.* **2015**, *46*, 102–105. [CrossRef]
5. Sekitani, T.; Nakajima, H.; Maeda, H.; Fukushima, T.; Aida, T.; Hata, K.; Someya, T. Stretchable active-matrix organic light-emitting diode display using printable elastic conductors. *Nat. Mater.* **2009**, *8*, 494–499. [CrossRef] [PubMed]
6. Kim, D.H.; Lu, N.; Ma, R.; Kim, Y.S.; Kim, R.H.; Wang, S.; Wu, J.; Won, S.M.; Tao, H.; Islam, A.; et al. Epidermal electronics. *Science* **2011**, *333*, 838–843. [CrossRef] [PubMed]
7. Kim, J.; Salvatore, G.A.; Araki, H.; Chiarelli, A.M.; Xie, Z.; Banks, A.; Sheng, X.; Liu, Y.; Lee, J.W.; Jang, K.I.; et al. Battery-free, stretchable optoelectronic systems for wireless optical characterization of the skin. *Sci. Adv.* **2016**, *2*, e1600418. [CrossRef] [PubMed]
8. Liu, Z.; Wang, X.; Qi, D.; Xu, C.; Yu, J.; Liu, Y.; Jiang, Y.; Liedberg, B.; Che, X. High-adhesion stretchable electrodes based on nanopile interlocking. *Adv. Mater.* **2016**, *29*, 1603382. [CrossRef] [PubMed]
9. Matsuhisa, N.; Kaltenbrunner, M.; Yokota, T.; Jinno, H.; Kuribara, K.; Sekitani, T.; Someya, T. Printable elastic conductors with a high conductivity for electronic textile applications. *Nat. Commun.* **2015**, *6*, 7461. [CrossRef] [PubMed]
10. Son, D.; Lee, J.; Qiao, S.; Ghaffari, R.; Kim, J.; Lee, J.E.; Song, C.; Kim, S.J.; Lee, D.J.; Jun, S.W.; et al. Multifunctional wearable devices for diagnosis and therapy of movement disorders. *Nat. Nanotechnol.* **2014**, *9*, 397–404. [CrossRef] [PubMed]
11. Xu, S.; Zhang, Y.; Jia, L.; Mathewson, K.E.; Jang, K.I.; Kim, J.; Fu, H.; Huang, X.; Chava, P.; Wang, R.; et al. Soft microfluidic assemblies of sensors, circuits, and radios for the skin. *Science* **2014**, *344*, 70–74. [CrossRef] [PubMed]
12. Minev, I.J.; Musienko, P.; Hirsch, A.; Barraud, Q.; Wenger, N.; Moraud, E.M.; Gandar, J.; Capogrosso, M.; Milekovic, T.; Asboth, L.; et al. Electronic dura mater for long-term multimodal neural interfaces. *Science* **2015**, *347*, 159–163. [CrossRef] [PubMed]
13. Graz, I.M.; Cotton, D.P.J.; Lacour, S.P. Extended cyclic uniaxial loading of stretchable gold thin-films on elastomeric substrates. *Appl. Phys. Lett.* **2009**, *94*, 071902. [CrossRef]
14. Graudejus, O.; Jia, Z.; Li, T.; Wagner, S. Size-dependent rupture strain of elastically stretchable metal. *Scr. Mater.* **2012**, *66*, 919–922. [CrossRef] [PubMed]
15. Adrega, T.; Lacour, S.P. Stretchable gold conductors embedded in PDMS and patterned by photolithography: Fabrication and electromechanical characterization. *J. Micromech. Microeng.* **2010**, *20*, 055025. [CrossRef]
16. Akogwua, O.; Kwabia, D.; Midturi, S.; Eleruja, M.; Babatope, B.; Soboyejoa, W.O. Large strain deformation and cracking of nano-scale gold films on PDMS substrate. *Mater. Sci. Eng. B* **2010**, *170*, 32–40. [CrossRef]
17. Jones, J.; Lacour, S.P.; Wagner, S. Stretchable wavy metal interconnects. *J. Vac. Sci. Technol. A Vac. Surf. Films* **2004**, *22*, 1723–1725. [CrossRef]
18. Lacour, S.P.; Jones, J.; Wagner, S.; Li, T.; Suo, Z. Stretchable interconnects for elastic electronic surfaces. *Proc. IEEE* **2005**, *93*, 1459–1467. [CrossRef]
19. Gray, D.S.; Tien, J.; Chen, S.C. High-conductivity elastomeric electronics. *Adv. Mater.* **2004**, *16*, 393–397. [CrossRef]
20. Gonzalez, M.; Axisa, F.; Bulcke, M.V.; Brosteaux, D.; Vandevelde, B.; Vanfleteren, J. Design of metal interconnects for stretchable electronic circuits. *Microelectron. Reliab.* **2008**, *48*, 825–832. [CrossRef]

21. Hsu, Y.Y.; Gonzalez, M.; Bossuyt, F.; Axisa, F.; Vanfleteren, J.; DeWolf, I. The effect of pitch on deformation behavior and the stretching-induced failure of a polymer-encapsulated stretchable circuit. *J. Micromech. Microeng.* **2010**, *20*, 075036. [CrossRef]
22. Löher, T.; Seckel, M.; Vieroth, R.; Dils, C.; Kallmayer, C.; Ostmann, A.; Aschenbrenner, R.; Reichl, H. Stretchable electronic systems: Realization and applications. In Proceedings of the 11th Electronic Packaging Technology Conference (EPTC2009), Singapore, 9–11 December 2009; pp. 893–898.
23. Koshi, T.; Iwase, E. Self-Healing metal wire using electric field trapping of metal nanoparticles. *Jpn. J. Appl. Phys.* **2015**, *54*, 06FP03. [CrossRef]
24. Koshi, T.; Iwase, E. Stretchable electronic device with repeat self-Healing ability of metal wire. In Proceedings of the 30th IEEE International Conference on Micro Electro Mechanical Systems (MEMS2017), Las Vegas, NV, USA, 22–26 January 2017; pp. 262–265.
25. Emry, R.D.; Povirk, G.L. Tensile behavior of free-standing gold films. Part II. Fine-grained films. *Acta Mater.* **2003**, *51*, 2067–2078. [CrossRef]
26. Zhang, S.; Sakane, M.; Nagasawa, T.; Kobayashi, K. Mechanical properties of copper thin films used in electronic devices. *Procedia Eng.* **2011**, *10*, 1497–1502. [CrossRef]
27. Jhonston, I.D.; McCluskey, D.K.; Tan, C.K.L.; Tracey, M.C. Mechanical characterization of bulk Sylgard 184 for microfluidics and microengineering. *J. Micromech. Microeng.* **2014**, *24*, 035017. [CrossRef]

micromachines

MDPI

Article

3D Shape Reconstruction of 3D Printed Transparent Microscopic Objects from Multiple Photographic Images Using Ultraviolet Illumination

Keishi Koyama [1], Masayuki Takakura [1], Taichi Furukawa [2] and Shoji Maruo [2,*]

[1] Graduate School of Engineering, Yokohama National University, 79-5 Tokiwadai, Hodogaya, Yokohama 240-8501, Japan; koyama-keishi-gx@ynu.jp (K.K.); takakura-masayuki-vw@ynu.jp (M.T.)
[2] Faculty of Engineering, Yokohama National University, 79-5 Tokiwadai, Hodogaya, Yokohama 240-8501, Japan; furukawa-taichi-xp@ynu.ac.jp
* Correspondence: maruo-shoji-rk@ynu.ac.jp; Tel.: +81-45-339-3880

Received: 11 May 2018; Accepted: 25 May 2018; Published: 27 May 2018

Abstract: We propose and demonstrate a simple, low-cost, three-dimensional (3D) shape acquisition method for transparent 3D printed microscopic objects. Our method uses ultraviolet (UV) illumination to obtain high-contrast silhouette images of transparent 3D printed polymer objects. Multiple silhouette images taken from different viewpoints make it possible to reconstruct the 3D shape of this transparent object. A 3D shape acquisition system consisting of a UV light-emitting diode, charge-coupled device camera and a rotation stage was constructed and used to successfully reconstruct the 3D shape of a transparent bunny model produced using micro-stereolithography. In addition, 3D printed pillar array models, with different diameters on the order of several hundred micrometers, were reconstructed. This method will be a promising tool for the 3D shape reconstruction of transparent 3D objects on both the micro- and macro-scale by changing the imaging lens.

Keywords: 3D shape reconstruction; shape from silhouette; 3D printing; additive manufacturing; micro-stereolithography; transparent object; photopolymer

1. Introduction

In recent years, various kinds of 3D printing technologies, from macro- to micro-scale devices, have been developed and widely used with a wide variety of materials including polymers, metals and ceramics [1–3]. To use 3D printed parts for final products, techniques for measuring the 3D shape of a 3D printed part are indispensable. X-ray computed tomography (CT) has been utilized as a powerful 3D shape measurement tool for 3D printed parts to date [4]. Recently, it has also been used to measure microscopic 3D printed parts [5]. However, because X-ray CT equipment is very expensive, it is not suitable as a convenient method for evaluating the 3D printed parts produced by the low-cost desktop 3D printers used by educators, hobbyists and professional designers.

On the other hand, several optical measurement methods including photogrammetry, structured light, shape from shading and shape from silhouette (SFS) have been developed as inexpensive techniques for acquiring the shape of a 3D object [6–10]. The SFS method, in particular, is a simple way to reconstruct a 3D shape of the target 3D object using multiple silhouette images captured from several directions. It has an advantage because it can be realized using only small and inexpensive pieces of equipment such as a camera, lighting device and rotary stage. Furthermore, the use of a zoom lens makes it possible to measure small 3D objects as small as 1 mm or less [10]. Therefore, it can be expected to be used for measuring the 3D printed microscopic objects produced by micro-scale 3D printing techniques such as single-photon micro- and two-photon stereolithography [11,12]. However, the conventional SFS method is difficult to use to measure transparent 3D objects such as the products of stereolithography and material jetting using ultraviolet (UV) curable polymers, because the silhouette of the transparent object includes the light that passed through the interior of the object in addition to its actual contour.

To overcome the above limitation of the conventional SFS method using visible light, we propose a novel method to acquire the shape of transparent 3D printed parts using UV illumination in this study. Most of the transparent 3D polymer parts produced by stereolithography absorb little visible light but absorb UV light strongly. Thus, it is possible to obtain high-contrast silhouette images using UV illumination. For this reason, even transparent 3D printed parts can be evaluated using the SFS method with UV light. Although there are some alternative methods for this, including local heating using infrared light [13] and the polarization of the reflected and emitted light [14], our method has advantages including the ability to capture silhouette images without a background subtraction process and a relatively high resolution as a result of the use of UV light.

We constructed a simple, low-cost 3D shape acquisition system using a UV light-emitting diode (LED), a UV-sensitive charge-coupled device (CCD) camera and a motorized rotation stage to demonstrate the usefulness of our proposed method. Using the optical system, we acquired the 3D shape of a bunny model as a case study. In addition, the accuracy of the 3D shape acquisition was evaluated by measuring an array of pillars with diameters ranging from 100–350 μm.

2. Materials and Methods

2.1. 3D Shape Acquisition Based on the Shape from Silhouette (SFS) Method

The first step in reconstructing the 3D shape of a 3D printed object using the SFS method is to acquire silhouette images of the target object from various directions. In the standard SFS method, a silhouette image is obtained by calculating the difference between an input image that includes the target object and a previously captured background image. In contrast, our SFS method using transmitted UV light illumination does not require background subtraction processes to capture silhouette images, because the background surrounding the target object has a uniform brightness and its contrast is enough high to binarize the silhouette images. Then, as shown in Figure 1, the binarized silhouette image on the image plane is back projected to the camera center to obtain a visual cone that includes the target object (a cube in Figure 1). Next, multiple visual cones are obtained from different viewpoints by positioning the camera around the object or rotating the object using a rotating stage. Finally, the common part (visual hull) of the visual cones obtained from each viewpoint is calculated. In principle, because the target object exists inside the visual hull, this visual hull can be used to acquire the 3D shape of the target object [9].

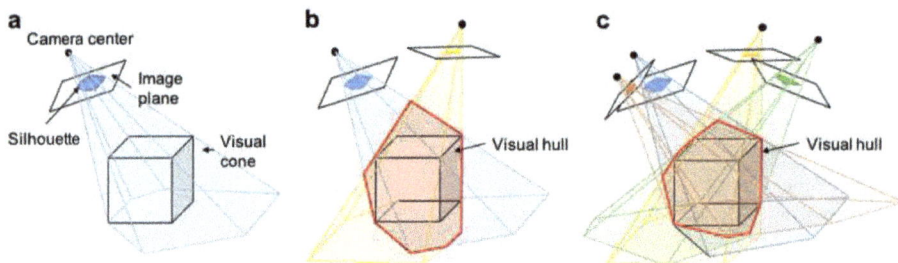

Figure 1. 3D shape reconstruction based on shape from silhouette (SFS) method. (**a**) visual cone obtained by back projection of silhouette image; (**b**) visual hull obtained by intersection of two visual cones; and, (**c**) 3D shape acquired by bounding geometry of resultant visual hull with multiple visual cones.

Our method employs the strategy reported by Atsushi et al. [10] for acquiring a 3D shape from silhouette images. In this method, a voxel-based 3D model is first used to represent the 3D shape of the target object in the SFS method. Then, the voxel-based 3D model is converted to a triangular mesh model using the marching cubes algorithm [15]. This triangular mesh model can be easily imported by commercial CAD software.

2.2. 3D Shape Acquisition System Based on SFS with Ultraviolet Light

To obtain an accurate, high-contrast silhouette image of a transparent 3D microscopic object, we constructed a 3D shape acquisition system using a UV LED (MBRL-CUV7530-2, Moritex Corp., Saitama, Japan, light-emitting area: 30×75 mm, emission peak wavelength: 365 nm), an imaging lens (MML1-ST150, Moritex Corp., numerical aperture: 0.038, magnification: $\times 1$, depth of field: 1.1 mm), a UV-sensitive CCD camera (XC-EU 50, Sony Corp., Tokyo, Japan) and a motorized rotation stage, as shown in Figure 2. The CCD camera is sensitive to light with wavelengths ranging from 300 nm to 420 nm. The silhouette images were captured at a resolution of 720×480 pixels. The CCD camera and attached lens were fixed at 60° angles from the horizontal plane. The rotation stage consisted of a stepping motor (AS 46 AAD, Oriental Motor Co., Ltd., Tokyo, Japan) and its controller (MSCTL 102, Suruga Seki Co., Ltd., Shizuoka, Japan).

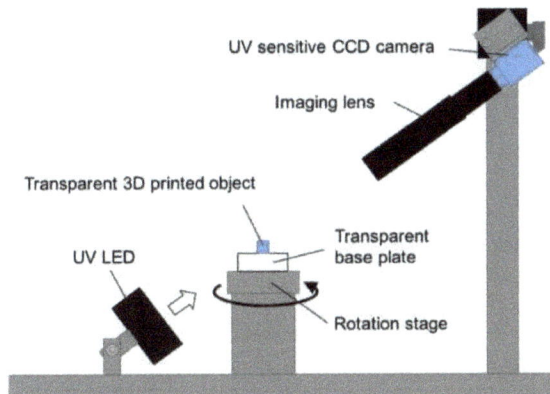

Figure 2. Optical setup for SFS using ultraviolet (UV) illumination. A silhouette image of the target object with UV illumination was captured by the UV sensitive charge-coupled device camera. Multiple silhouette images with different viewpoints were obtained by rotating the rotation stage at 10° intervals.

To reconstruct the 3D shape from the acquired silhouette images, in the SFS method, it was necessary to calibrate the internal and external parameters of the camera. Our experiments used a calibration method developed by Lavest et al. [16] to determine parameters such as the distance between the camera and the object and the relationship between the acquired image and the real-world coordinates. A 2 mm square cube made by cutting was the target object used to calibrate the camera.

After camera calibration, a transparent 3D printed microscopic part was placed in the center of the rotation stage and its silhouette images from different viewpoints were captured by rotating the stage at 10° intervals. The 3D shape of the 3D printed part was reconstructed using the captured silhouette images with a reconstruction program developed by Atsushi et al. [10].

2.3. Micro-Stereolithography Systems and Photocurable Resins

In our experiments, two types of laboratory-made micro-stereolithography systems were used to make 3D printed micro-parts. One was a bottom-up system (free-surface method) based on single-photon polymerization using a He-Cd laser (IK5551R-F, Kimmon Koha Co., Ltd., Tokyo, Japan, wavelength: 325 nm). The laser spot size of this system is approximately 12 μm. The minimum layer thickness for 3D printing is 30 μm because of the viscosity of the resin. Therefore, this laser was suitable for making millimeter-sized 3D objects such as microchannels, scaffolds and energy harvesters [11,17,18]. Therefore, this system was used to make a bunny model (size: 1.2 × 0.8 × 1.1 mm) using a commercial epoxy-based photocurable resin (TSR-883, CMET Inc., Yokohama, Japan) as an example of a transparent complex 3D object. The other laboratory-made micro-stereolithography system used was a top-down system (constrained-surface method) based on single-photon polymerization using a blue laser diode (Cobolt 06-MLD, Cobolt AB, Solna, Sweden, wavelength: 405 nm). In this system, the blue laser beam is collimated and introduced into a Galvano scanner (GM-1010, Canon Inc., Tokyo, Japan) and focused by an objective lens (numerical aperture: 0.1). Both the focal spot size and minimum layer thickness of this system are 5 μm. Therefore, this system was used to create finer structures with higher resolutions compared to those created by the bottom-up system. The resin used in the bottom-up system is not suitable for the system using a blue laser beam because of its absorption spectrum. Therefore, we prepared a laboratory-made photocurable resin containing an acrylate monomer (SR399, Sartomer Inc., Exton, PA, USA, 95.1 wt %), a photoinitiator (Diphenyl(2,4,6-trimethylbenzoyl)phosphine oxide, Sigma-Aldrich, St. Louis, MO, USA, 1.0 wt %), an inhibitor (2-tert-Butyl-4-methylphenol, Sigma-Aldrich, St. Louis, MO, USA, 2.9 wt %) and a UV absorber (2-(5-Chloro-2-benzotriazolyl)-6-tert-butyl-p-cresol, Tokyo Chemical Industry Co. Ltd., Tokyo, Japan, 1.0 wt %). Using the top-down system with the laboratory-made acrylate-based photocurable resin, we fabricated a pillar array model that had four pillars with different diameters.

3. Results and Discussion

3.1. Transmission Spectrum of Photopolymer

To measure the transmittance values of the two kinds of cured resins, thin films with a thickness of approximately 200 μm were prepared by curing both resins with a UV lamp. The transmission spectra of the thin films were measured using a UV-visible spectrometer (UV-1700, SHIMADZU Corp., Kyoto, Japan) (Figure 3). The transmittance values of the acrylate-based and epoxy-based resins at the UV LED emission peak wavelength (365 nm) were 0.1% and 46.3%, respectively. Therefore, the acrylate-based resin was considered suitable for obtaining high-contrast silhouette images using the UV LED. Although the UV absorption of the epoxy-based resin was lower than that of the acrylate-based resin, it could also be used to obtain a sufficient number of high-contrast silhouette images after proper binarization without a background subtraction process, as shown in the following experiments.

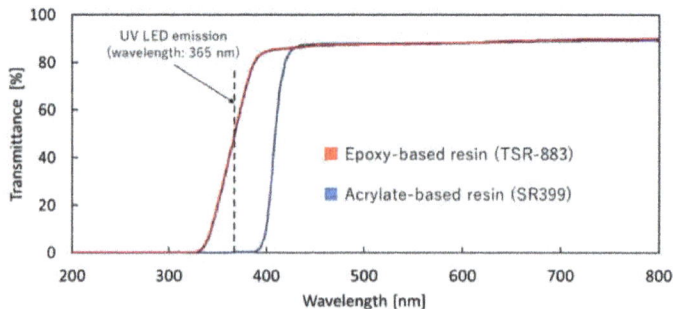

Figure 3. Transmission spectra of two cured resins: a commercial epoxy-based resin (TSR-883) and a laboratory-made acrylate-based resin containing SR399.

3.2. 3D Shape Acquisition of Transparent 3D Printed Objects

To demonstrate the usefulness of our proposed method, we acquired the 3D shape of a miniature bunny model produced using micro-stereolithography based on the bottom-up system. Figure 4 compares silhouette images obtained using UV and visible light. In these experiments, the transparent bunny model made from the epoxy-based resin was used as the target object (Figure 4a). To obtain a silhouette image with visible light, we replaced the UV LED with a halogen fiber light source (LG-PS2, Olympus Corp., Tokyo, Japan) using the optical setup shown in Figure 2. As shown in Figure 4b, the resulting silhouette image was a blocky, gray photograph caused by transmission and refraction from the transparent object. It was not suitable for obtaining a correct binarized silhouette image without artificial voids. On the other hand, the silhouette image obtained using UV transmitted illumination was a substantially uniform, black photograph because most of the UV light was absorbed. Therefore, a correct silhouette of the transparent target object could be obtained after proper binarization of the silhouette image. This shows the advantage of using UV transmitted illumination for capturing correct, high-contrast silhouette images.

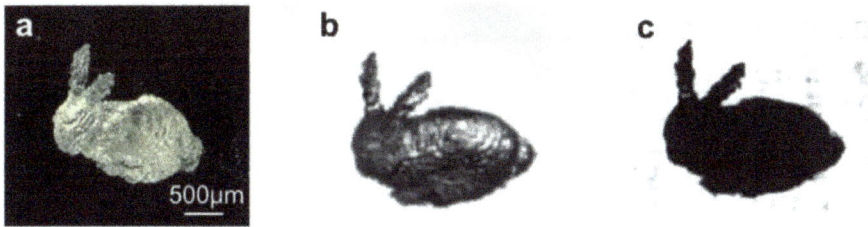

Figure 4. Comparison of silhouette images of a 3D printed bunny model using visible and UV light. (**a**) optical microscope image of an epoxy-based resin model of a bunny, produced using micro-stereolithography; (**b**) silhouette image obtained with visible transmitted light; and, (**c**) silhouette image obtained with UV transmitted light.

The 3D shape of the bunny model was reconstructed using the SFS method with visible and UV light. Figure 5 shows the reconstructed triangular mesh models. As Figure 5a shows, some portions of the bunny model could not be reconstructed using visible light. On the other hand, all portions of the bunny model were completely reconstructed using UV light. This was because the use of high-contrast silhouette images with slightly uneven brightness levels made it possible to obtain the correct visual hull of the target object.

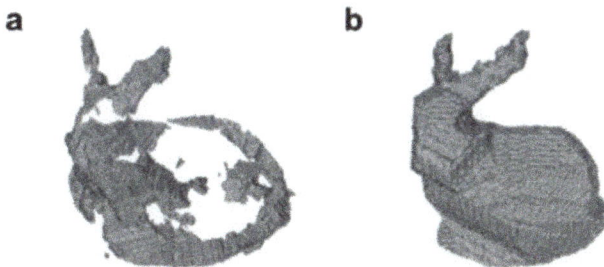

Figure 5. 3D shape reconstruction of a bunny model using (**a**) visible light and (**b**) UV transmitted light illumination.

3.3. Evaluating the Accuracy of 3D Shape Acquisition Using the Pillar Array Model

To evaluate the accuracy of the 3D shape acquisition system, a pillar array model (Figure 6a) containing pillars of different diameters was fabricated using micro-stereolithography based on the top-down system. The diameter of each of the actual pillars was measured using an optical microscope; these results are summarized in Table 1. Figure 6b shows the silhouette image of the pillar array model captured using UV transmitted illumination. All of the pillars were captured with high contrast. The 3D shape of the pillar array model was reconstructed using multiple silhouette images taken from different viewpoints (Figure 6c). Although the smallest pillar was slightly distorted, all the pillars were reconstructed. The average diameter of each of the reconstructed pillars was calculated using a cross section at half the height of each pillar. In this calculation, we used the average of an inscribed circle and a circumscribed circle for each pillar, as shown in Figure 6c. The averaged diameters of these reconstructed pillars are summarized in Table 1. In these results, the difference between the actual and averaged diameters of the reconstructed pillars ranged from 1–20 μm.

There are some parameters that could be used to reduce the errors in the reconstructed 3D shape. In the SFS method, we used a pixel size of 11.7 μm to represent the voxel-based model. The reconstructed 3D model could be smoother and finer if a smaller voxel size was used. The number of CCD camera elements also affected the quality of the silhouette images. Using a higher resolution camera could also reduce the minimum pixel size of the silhouette images and make the visual cone more precise. Additionally, the magnification and depth of field of the imaging lens are important parameters. Since there is a trade-off relationship between the magnification and depth of field, observing 3D microscopic objects using an optical microscope is an intrinsic problem. As shown in Figure 6b, the silhouette image of the pillars has a high contrast but the focus is blurry because of the limited depth of field. To overcome these problems, we could use an image fusion technique for a sequence of images taken by changing the position of the focus along the optical axis. This technique would provide sharp silhouettes even under a high magnification [19].

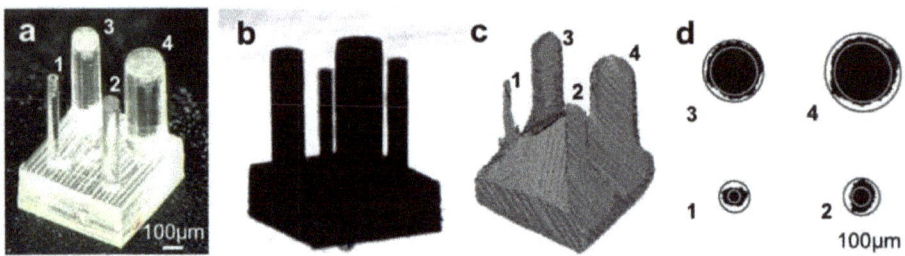

Figure 6. 3D shape reconstruction of a 3D printed pillar array model made of acrylate-based resin. (**a**) optical microscope image; (**b**) silhouette obtained using UV illumination; (**c**) reconstruction of the pillar array model using a shape from silhouette method; and (**d**) a cross section at half the height of each reconstructed pillar.

Table 1. Diameter of each pillar of a 3D printed pillar array model.

Pillar Number	Actual Pillar Diameter Measured by an Optical Microscope	Averaged Pillar Diameter Estimated by SFS Method
1	106 μm	94 μm
2	152 μm	151 μm
3	248 μm	268 μm
4	347 μm	333 μm

4. Conclusions

We demonstrated a simple and low-cost 3D shape acquisition method for transparent 3D printed microscopic objects. This method employed highly UV-absorbent 3D printed polymer objects to obtain high-contrast silhouette images of transparent 3D objects using UV transmitted illumination. Multiple silhouette images taken from different viewpoints made it possible to reconstruct the 3D shapes of the transparent 3D printed objects using the SFS method, with a 3D shape acquisition system constructed using a UV LED, a CCD camera and a rotation stage. A bunny model as small as 1 mm was successfully reconstructed with this system using an imaging lens with a 1× magnification. By changing the imaging lens, this system could be applicable to macro- and micro-scale models. In addition, transparent 3D printed models made from glass as well as polymer [20,21] could be observed using this method. Therefore, this method could be an inexpensive and useful tool for a 3D scanner and a way to inspect the appearance of transparent 3D objects without the need for time-consuming pre- and post-processing techniques.

Author Contributions: K.K. and T.F. performed the experiments. K.K. and M.T. calculated and evaluated the 3D shapes of the target objects. K.K. and S.M. wrote the paper. S.M. supervised the research.

Funding: This work was supported by the Cross-ministerial Strategic Innovation Promotion Program (SIP) of the New Energy and Industrial Technology Development Organization (NEDO).

Acknowledgments: We thank Takashi Maekawa for the useful discussion on 3D shape reconstruction and for providing software for the SFS method. We also thank CMET Inc. for providing the photocurable resin (TSR-883).

Conflicts of Interest: The authors declare no conflict of interest.

References

1. Ligon, S.C.; Liska, R.; Stampfl, J.; Gurr, M.; Mülhaupt, R. Polymers for 3D printing and customized additive manufacturing. *Chem. Rev.* **2017**, *117*, 10212–10290. [CrossRef] [PubMed]
2. William, E.F. Metal additive manufacturing: A review. *J. Mater. Eng. Perform.* **2014**, *23*, 1917–1928. [CrossRef]
3. Deckers, J.; Vleugels, J.; Kruth, J.P. Additive manufacturing of ceramics: A review. *J. Ceram. Sci. Technol.* **2014**, *5*, 245–260. [CrossRef]
4. Thompson, A.; Maskery, I.; Leach, R.K. X-ray computed tomography for additive manufacturing: A review. *Meas. Sci. Technol.* **2016**, *27*, 1–17. [CrossRef]
5. Saha, S.K.; Oakdale, J.S.; Cuadra, J.A.; Divin, C.; Ye, J.; Forien, J.B.; Aji, L.B.; Biener, J.; Smith, W.L. Radiopaque resists for two-photon lithography to enable submicron 3D imaging of polymer parts via X-ray computed tomography. *ACS Appl. Mater. Interfaces* **2018**, *10*, 1164–1172. [CrossRef] [PubMed]
6. Chen, F.; Brown, G.M.; Song, M. Overview of three-dimensional shape measurement using optical methods. *Opt. Eng.* **2000**, *39*, 10–22. [CrossRef]
7. Moons, T.; Gool, L.V.; Vergauwen, M. 3D Reconstruction from multiple images part 1: Principles. *Comput. Graph. Vis.* **2008**, *4*, 287–398. [CrossRef]
8. Zhang, R.; Tsai, P.S.; Cryer, J.E.; Shah, M. Shape-from-shading: A survey. *IEEE Trans. Pattern Anal. Mach. Intell.* **1999**, *21*, 690–706. [CrossRef]
9. Cheung, G.; Baker, S.; Kanade, T. Shape-from-silhouette across time part I: Theory and algorithms. *Int. J. Comput. Vis.* **2005**, *62*, 221–247. [CrossRef]
10. Atsushi, K.; Sueyasu, H.; Funayama, Y.; Maekawa, T. System for reconstruction of three-dimensional micro objects from multiple photographic images. *Comput. Aided Des.* **2011**, *43*, 1045–1055. [CrossRef]
11. Monri, K.; Maruo, S. Three-dimensional ceramic molding based on microstereolithography for the production of piezoelectric energy harvesters. *Sens. Actuators A* **2013**, *200*, 31–36. [CrossRef]
12. Maruo, S.; Fourkas, J.T. Recent progress in multiphoton microfabrication. *Laser Photon. Rev.* **2008**, *2*, 100–111. [CrossRef]
13. Gonen, E.; Olivier, A.; Fabrice, M.; Sanchez, L.A.; David, F.; Teoman, N.A.; Frederic, T.; Aytul, E. Scanning from heating: 3D shape estimation of transparent objects from local surface heating. *Opt. Express* **2009**, *17*, 57–68. [CrossRef]

14. Miyazaki, D.; Saito, M.; Sato, Y.; Ikeuchi, K. Determining surface orientations of transparent objects based on polarization degrees in visible and infrared wavelengths. *J. Opt. Soc. Am.* **2002**, *19*, 687–705. [CrossRef]

15. William, E.L.; Cline, H.E. Marching cubes: A high resolution 3D surface construction algorithm. *ACM Siggraph Comput. Graph.* **1987**, *21*, 163–169. [CrossRef]

16. Lavest, J.M.; Rives, G.; Dhome, M. Three-dimensional Reconstruction by Zooming. *IEEE J. Robot. Autom.* **1993**, *9*, 196–207. [CrossRef]

17. Inada, M.; Hiratsuka, D.; Tatami, J.; Maruo, S. Fabrication of three-dimensional transparent SiO_2 microstructures by microstereolithographic molding. *Jpn. J. Appl. Phys.* **2009**, *48*, 06FK01. [CrossRef]

18. Torii, T.; Inada, M.; Maruo, S. Three-dimensional molding based on microstereolithography using beta-tricalcium phosphate slurry for the production of bio ceramic scaffolds. *Jpn. J. Appl. Phys.* **2011**, *50*, 06GL15. [CrossRef]

19. Gallo, A.; Muzzupappa, M.; Bruno, F. 3D reconstruction of small sized objects from a sequence of multi-focused images. *J. Cult. Herit.* **2014**, *15*, 173–182. [CrossRef]

20. Kotz, F.; Arnold, K.; Bauer, W.; Schild, D.; Keller, N.; Sachsenheimer, K.; Nargang, T.M.; Richter, C.; Helmer, D.; Rapp, B.E. Three-dimensional printing of transparent fused silica glass. *Nature* **2017**, *544*, 337–339. [CrossRef] [PubMed]

21. Klein, J.; Stern, M.; Franchin, G.; Kayser, M.; Inamura, C.; Dave, S.; Weaver, J.C.; Houk, P.; Colombo, P.; Yang, M.; et al. Additive manufacturing of optically transparent glass. *3D Print. Addit. Manuf.* **2015**, *2*, 92–105. [CrossRef]

micromachines

MDPI

Article

Effect of Heat Accumulation on Femtosecond Laser Reductive Sintering of Mixed CuO/NiO Nanoparticles

Mizue Mizoshiri [1],* [ID], Kenta Nishitani [2] and Seiichi Hata [2]

1   Department of Mechanical Engineering, Nagaoka University of Technology, 1603-1, Kamitomioka, Nagaoka, Niigata 940-2188, Japan
2   Department of Micro-Nano Mechanical Science and Engineering, Nagoya University, Furo-cho, Chikusa, Nagoya, Aichi 464-8603, Japan; nishitani.kenta@a.mbox.nagoya-u.ac.jp (K.N.); hata@mech.nagoya-u.ac.jp (S.H.)
*   Correspondence: mizoshiri@mech.nagaokaut.ac.jp; Tel.: +81-258-479-765

Received: 11 May 2018; Accepted: 24 May 2018; Published: 28 May 2018

Abstract: Direct laser-writing techniques have attracted attention for their use in two- and three-dimensional printing technologies. In this article, we report on a micropatterning process that uses femtosecond laser reductive sintering of mixed CuO/NiO nanoparticles. The writing speed, laser fluence, and incident total energy were varied to investigate the influence of heat accumulation on the micropatterns formed by these materials. Heat accumulation and the thermal history of the laser irradiation process significantly affected the material composition and the thermoelectric properties of the fabricated micropatterns. Short laser irradiation durations and high laser fluences decrease the amount of metal oxide in the micropatterns. Selective fabrication of p-type and n-type thermoelectric micropatterns was demonstrated to be possible with control of the reduction and reoxidization reactions through the control of writing speed and total irradiation energy.

Keywords: direct writing; femtosecond laser; reductive sintering; thermoelectric film; Cu-Ni alloy; micropatterns; printing

1. Introduction

Direct laser-writing techniques have attracted attention in printing technologies. For example, three-dimensional (3D) printing, which is known as additive manufacturing, has been used in the fabrication of 3D bulk metal structures. In the process, raw metal powders are sintered and melted with heat from a laser in an inert atmosphere such as a vacuum or an inert gas. To date, structures made of many metals and alloys, such as Cu, Fe-Ni, and Ti-6Al-4V, have been achieved by selectively sintering and melting powders of the materials [1–4]. However, it is difficult to scale these processes for producing metal microstructures because fine metal powders oxidize too easily.

To fabricate 2D and 3D metal microstructures, metals and alloys can be printed from metal nanoparticle (NP) inks and metal oxide NP solutions. For example, metal NP inks with Au, Ag, or Cu NPs mixed with dispersants have been selectively sintered using lasers in ambient atmosphere [5–7]. Neodymium-doped yttrium aluminum garnet (Nd/YAG) lasers operating with the wavelength of 532 nm are particularly effective for Cu sintering because Cu exhibits higher absorbance for this wavelength than it does for infrared light. Flexible displays composed of such metal electrodes have been fabricated using the printing techniques.

Metal oxide NP solutions with NPs of CuO, Cu_2O, and NiO, a reducing agent, and a dispersant, have been used for laser reductive sintering [8–11]. A CuO NP solution composed of CuO NPs, ethylene glycol (EG) as the reductant, and polyvinylpyrrolidone (PVP) as the dispersant, was reduced and sintered using continuous-wave and nanosecond lasers at a wavelength of 1070 nm [8]. The advantage of the process is that the CuO NP solution considerably absorbs the laser light since the band gap of CuO is 1.2 eV (wavelength 1033 nm). Two-dimensional Cu micropatterns have been formed on glass substrates and polyimide films using reductive sintering [8]. In another work, Cu_2O NP solutions were prepared by mixing of Cu_2O NPs, 2-propanol, and PVP. When a laser beam is focused onto the Cu_2O NP solutions, formic acid is generated by the thermal reaction of 2-propanol and PVP. Then, the formic acid reduces the Cu_2O to Cu [9]. Ni micropatterns have also been fabricated by laser reductive sintering of NiO NPs [10,11]. In this process, solutions of NiO NPs and toluene are reduced by irradiation with a 514.5-nm continuous wave laser to fabricate 2D and Ni patterns.

We have also developed a process for femtosecond laser reductive sintering of metal oxide NPs [12–14]. Femtosecond laser pulses are effective for controlling the reduction and reoxidization of the micropatterns. For example, Cu-rich and Cu_2O-rich micropatterns can be formed selectively by controlling laser irradiation conditions such as the writing speed [12]. We have used this process to fabricate Cu/Cu_2O composite micro-thermistors. Furthermore, NP solutions with the mixtures of NiO/Cr and CuO/NiO have enabled us to form Cu-Ni alloys and Ni/Cr-O composite micropatterns, respectively [13,14]. Ni/Cr-O microgears were successfully fabricated, and they could be moved by controlling an external magnetic field. Cu-Ni and Cu_2O/NiO micropatterns exhibit n-type and p-type thermoelectric properties, respectively [14]. In addition, we applied the selective micropatterning process to demonstrate the performance of thermocouples. The selective fabrication of Cu-Ni and Cu_2O/NiO micropatterns is possible by controlling the laser-writing speed to be as high as 1–20 mm/s, which can lead to problems in the fabrication process. The effects of heat accumulation on the reduction and reoxidization reactions are complex because the balance of the reduction and reoxidization reactions is determined by the chemical potentials and activation energies of the metals (Cu, Ni) and O. Micropatterns are typically formed by raster scanning of a focused laser beam. Therefore, such high writing speeds are not ideal since the scale of the micropatterns is limited by the significant effects to the material around the edges of the laser-irradiated area.

This article reports our investigation of the effects of heat accumulation on micropatterns formed through femtosecond laser reductive sintering of CuO/NiO mixed NPs. The details of the patterning properties were evaluated at writing speeds including below 1 mm/s. The crystal structures and metal oxide composites of the fabricated micropatterns were subsequently examined. Then, the Seebeck coefficients of the micropatterns were calculated. Finally, we fabricated thermocouples to demonstrate the process's effectiveness for precisely fabricating fine micropatterns with controllable thermoelectric properties.

2. Materials and Methods

CuO/NiO NP solution was prepared by mixing CuO NPs (Sigma Aldrich, St. Louis, MO, USA, diameter < 50 nm), NiO NPs (Sigma Aldrich, St. Louis, MO, USA, diameter < 50 nm), PVP (Sigma Aldrich, St. Louis, MO, USA, M_w~10,000), and EG (Sigma Aldrich, St. Louis, MO, USA) using ultrasonic agitation. The concentrations of CuO NPs, NiO NPs, PVP, and EG were 36.9 wt %, 23.1 wt %, 13 wt %, and 27 wt %, respectively. Then, the CuO/NiO NP solution was spin-coated on 1-mm-thick glass substrates. The thickness of the coated film was ~10 μm. Direct writing was subsequently performed in air using a femtosecond laser-writing system (Photonic Professional GT, Nanoscribe GmbH, Eggenstein-Leopoldshafen, Germany). Femtosecond laser pulses operating at a wavelength of 780 nm, repetition rate of 80 MHz, and pulse duration of 120 fs were focused onto the surface of the films using an objective lens with numerical aperture of 0.75. The focused beam diameter was ~1.3 μm. The laser polarization was linear. Micropatterns were formed by scanning the focused laser pulses using an x-y mechanical stage. The writing speed was varied in the range of 100–5000 μm/s. The micropatterns were written by raster scanning of the laser focal spot at a pitch of 10 μm, which was determined by considering the minimum line width of ~10 μm. Finally, residual non-irradiated NPs were removed by rinsing the substrates in EG and ethanol.

The morphology of the micropatterns was observed using a scanning electron microscope (SEM, Hitachi High Technologies, Tokyo, Japan, TM3030Plus). These observations were performed in the SEM's low vacuum mode without coating of the electrical thin films to reduce their electrical charge. The crystal structures of the fabricated micropatterns were examined using an imaging plate X-ray diffraction (XRD) apparatus (Rigaku, Tokyo, Japan, Rint Rapid-S diffractometer) with a collimated beam diameter of 0.3 mm. The incident angle was fixed to 20°. The oxide materials included in the micropatterns, such as CuO, Cu_2O, and NiO, were evaluated via Raman spectrometry using a 532-nm laser.

The Seebeck coefficient S of each micropattern was estimated by measuring both the temperature difference ΔT between the fabricated micropatterns and the voltage V generated by that temperature difference. Infrared thermography (Nippon Avionics, Tokyo, Japan, Thermo Shot F30) was used to assess the temperature difference. The voltage generated was measured with a multimeter (Keysight Technology, Santa Rosa, CA, USA, Truevolt series 34465A). The Seebeck coefficient S was defined as $V/\Delta T$.

3. Results

In preliminary tests, the crystal structures of the fabricated micropatterns were clarified with XRD analysis. The oxide compositions of the micropatterns were also measured using Raman spectroscopy. Then, we measured the micropattern responses with various writing conditions such as the writing speed, incident total energy, and laser fluence to evaluate effects of heat accumulation on the fabricated micropatterns.

3.1. Micropatterns at Various Writing Speed

First, the morphologies of the fabricated micropatterns were observed using SEM images. Figure 1a–e show the surface morphologies of the micropatterns fabricated at a laser fluence of 0.059 J/cm^2 and writing speeds of 100, 500, 1000, 3000, and 5000 μm/s, respectively. Lines from the laser scanning appeared in the form of grooves at the low writing speeds of 100–1000 μm/s. The surfaces were partially melted at these low writing. We consider that irradiating laser pulses with a high repetition rate induced the heat accumulation that ablated the material at low writing speeds However, micropatterns with somewhat uniform surfaces were formed at the writing speed of 5000 μm/s. In this case, the raw NPs seem to have been sintered instead of melted.

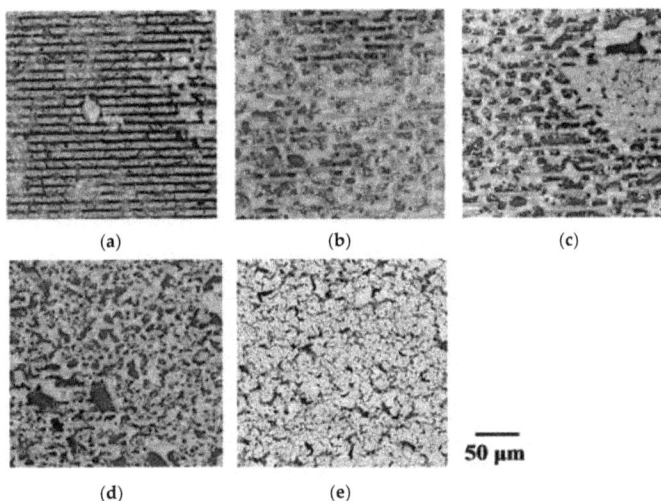

Figure 1. SEM images of micropatterns fabricated at writing speeds of (**a**) 100 μm/s, (**b**) 500 μm/s, (**c**) 1000 μm/s, (**d**) 3000 μm/s, and (**e**) 5000 μm/s, respectively.

Figure 2 shows XRD spectra recorded from the micropatterns. Metal or metal oxide composite micropatterns were obtained, depending on the writing speed. With the fast writing speed of 5000 μm/s, XRD intensity peaks corresponding to the oxides were weak. On the other hand, intense XRD peaks corresponding to Cu_2O and NiO appeared with low writing speeds of 100–3000 μm/s. CuO was clearly generated only at the writing speed of 100 μm/s. The broad spectra were observed between the peaks of Cu and Ni in Figure 2a. In addition, the peak shift was also observed at 43°–44° in the micropatterns written at 5000 μm/s in Figure 2b. These results suggest the possibility of the generation of Cu-Ni alloy or metal oxides.

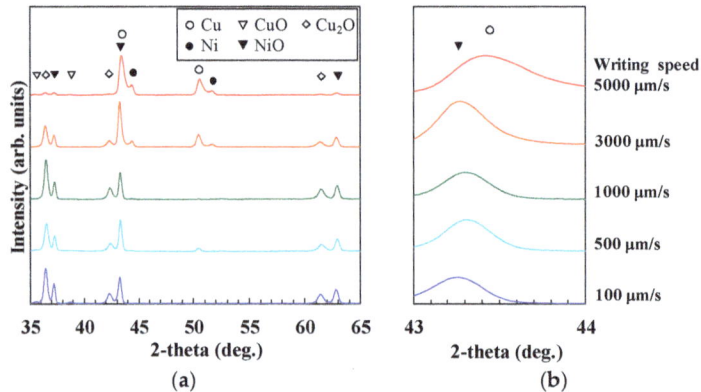

Figure 2. (**a**) X-ray diffraction (XRD) spectra of micropatterns, (**b**) enlargement of these spectra from 43–44 degrees.

The separations of the diffraction peaks for NiO and Cu at ~43.3° are difficult to distinguish. To examine the composition ratio of the metal and metal oxide in the micropatterns, Raman spectra of the fabricated micropatterns were evaluated, and the evaluation results are shown in Figure 3. No peak appeared with the writing speed of 5000 μm/s, suggesting that metal oxides were not included in the micropatterns. However, obvious peaks corresponding to NiO appeared at the writing speeds of 500 and 1000 μm/s. By considering the XRD and Raman spectra of the micropatterns together, we found that the XRD peak shift at 43°–44° in the micropatterns written at 5000 μm/s indicates the generation of Cu-Ni alloy because the Raman spectra show that the micropatterns did not include metal oxides.

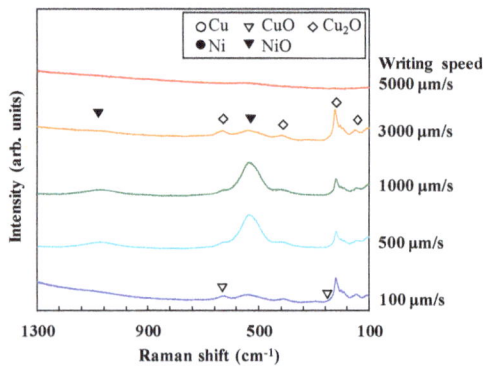

Figure 3. Raman shifts of the fabricated micropatterns.

The Seebeck coefficients of the micropatterns are listed in Table 1. p-type micropatterns were obtained with the low writing speeds of 100–500 µm/s. This result is consistent with XRD and Raman spectra results that indicate the generation of Cu_2O and NiO, which is a p-type material. n-type micropatterns were obtained at the fast writing speed of 5000 µm/s. XRD and Raman spectra results indicate the generation of low amounts of metal oxides such as Cu_2O, CuO, and NiO, and high amounts of Cu-Ni alloy in the n-type thermoelectric material. The n-type thermoelectric properties generated with the writing speed of 5000 µm/s are owing to the presence of Cu-Ni alloys.

Table 1. Seebeck coefficients of micropatterns printed at various writing speeds.

Writing Speed (µm/s)	Seebeck Coefficient (µV/K)	Thermoelectric Type
100	2.7×10^2	p-type
500	9.7×10^2	p-type
1000	0	-
3000	0	-
5000	−29	n-type

3.2. Micropatterns Formed with Various Laser Fluences

Next, the effects of the laser fluence on the micropatterns were investigated while the total incident energy kept constant. The writing speed was decided depending on the laser fluence to maintain constant total incident energy across these trials. Table 2 lists the laser irradiation conditions.

Table 2. Laser irradiation conditions at constant total irradiation energy.

Laser Fluence (J/cm^2)	Writing Speed (µm/s)
0.012	1000
0.024	2000
0.035	3000
0.047	4000
0.059	5000

Figure 4a–d show SEM images of the surface morphology of the micropatterns fabricated with laser fluence 0.024–0.059 J/cm^2. The laser fluence of 0.012 J/cm^2 did not form any micropatterns because they were washed from the substrate during the rinsing step that removed non-irradiated NPs. Compared to the different incident total energies in Figure 1, all the micropatterns have similar surfaces. They seem to have been formed by the sintering of the NPs without any obvious melting. This result indicates that the total incident energy determined the surface morphology of the micropatterns.

To evaluate the proportions of metals and metal oxides included in the micropatterns, the intensity ratios of the XRD spectra $I(Cu_{(111)}/CuO_{(111)})$ for generating Cu, $I(Cu_2O_{(111)}/CuO_{(111)})$ for generating Cu_2O, $I(Cu_{(111)}/Cu_2O_{(111)})$ for evaluating the degree of CuO reduction, and $I(Ni_{(111)}/NiO_{(111)})$ for generating Ni, were calculated. Figure 5a–d show the intensity ratios for each micropattern. The generation of both Cu and Cu_2O increased with increasing laser fluence, as shown in Figure 4a,b. In addition, Cu generation increased with increasing laser fluence. These results indicate that high laser fluence is important to reduce the CuO/NiO NPs entirely. High and low laser fluence levels are effective for generation of Cu and Cu_2O, respectively.

However, Ni generation in the micropatterns at the laser fluence of 0.059 J/cm^2 was less than that achieved with the fluence of 0.047 J/cm^2. This difference suggests that the Ni was reoxidized at the laser fluence of 0.059 J/cm^2 because Ni is oxidized more easily than Cu. Therefore, the high generation of Cu may be induced by the contributing of Ni acting as a reductant.

The Seebeck coefficients of the micropatterns are listed in Table 3. All the micropatterns exhibited n-type thermoelectric properties. The largest negative value was obtained at the laser fluence of 0.059 J/cm^2, that is consistent with the small amount of generation of the p-type oxides such as Cu_2O and NiO.

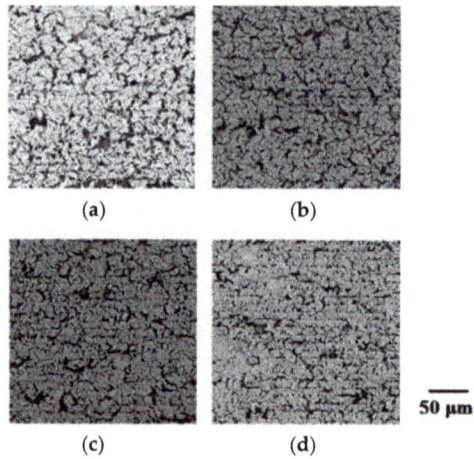

Figure 4. SEM images of the fabricated micropatterns at laser fluences of (**a**) 0.059 J/cm^2, (**b**) 0.047 J/cm^2 μm/s, (**c**) 0.035 J/cm^2, and (**d**) 0.024 J/cm^2.

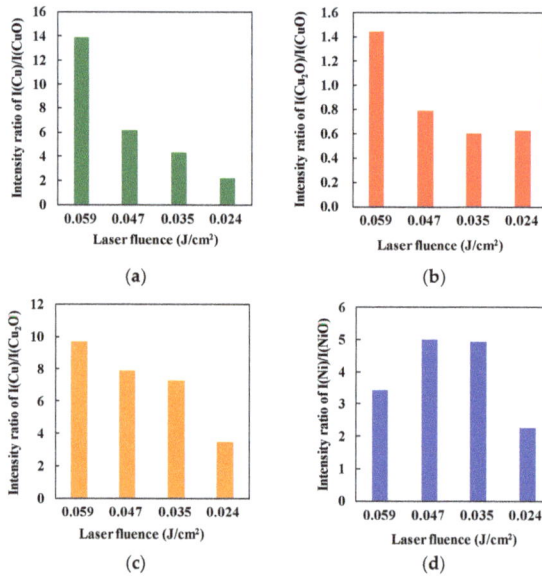

Figure 5. Intensity ratios of (**a**) I(Cu)/I(CuO), (**b**) I(Cu$_2$O)/I(CuO), (**c**) I(Cu)/I(Cu$_2$O), and (**d**) I(Ni)/I(NiO).

Table 3. Seebeck coefficients of the micropatterns at constant total irradiation energy and various writing speeds.

Laser Fluence (J/cm^2)	Writing Speed (μm/s)	Seebeck Coefficient (μV/K)	Thermoelectric Type
0.012	1000	No pattern	No pattern
0.024	2000	−11	n-type
0.035	3000	−18	n-type
0.047	4000	−19	n-type
0.059	5000	−29	n-type

3.3. Effect of Heat Accumulation on Micropatterning

To evaluate the effect of heat accumulation on the micropatterning, single- and double-exposed micropatterns were fabricated. The laser fluences were chosen as 0.059 J/cm^2 for single exposure and 0.030 J/cm^2 for double exposure. The writing speed was fixed at 500 μm/s. In the double exposure, firstly the micropattern was written by raster scanning. Then, the second scanning was performed on the first exposed area. The duration between the end of the first exposure and the start of the second exposure was several few seconds, which was expected to be enough to cool the heated materials at the first exposure. Figure 6a,b show SEM images of the surface morphology of these micropatterns. Deep grooves appeared with single writing at high laser fluence. These results indicate that the high laser fluence caused rapid heating and ablation where the laser was irradiated the material. XRD spectra of the micropatterns are shown in Figure 6c. The generation of metal oxide was higher with double exposure at 0.030 J/cm^2 than that with single writing at 0.059 J/cm^2. Compared to the case of double exposure with a low laser fluence, the maximum temperature is high in the case of single writing with a high laser fluence. Therefore, the reduction and sintering of the metal oxide NPs seems to be caused in short duration, preventing the reoxidation of the reduced metal NPs.

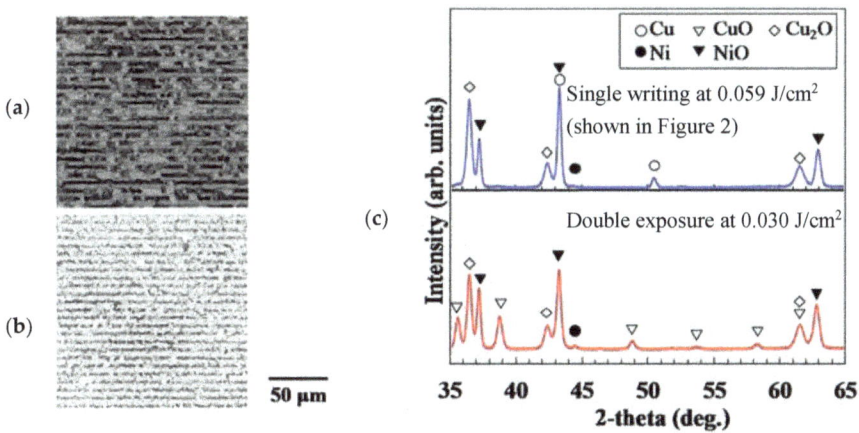

Figure 6. SEM images of the surfaces of micropatterns fabricated with (**a**) single exposure at 0.059 J/cm^2 and (**b**) double exposure at 0.030 J/cm^2; (**c**) XRD spectra of the micropatterns.

The Seebeck coefficient of the micropatterns was also estimated. With double exposure at 0.030 J/cm^2, the Seebeck coefficient of the micropatterns was 57.9 μV/K, which was smaller than that observed with single writing at 0.059 J/cm^2. These results are listed in Table 3. This small Seebeck coefficient was induced by the generation of CuO because the temperature gradient that contributes to the thermoelectric voltage is decreased by the CuO. Therefore, the open-circuit voltage, through which we estimate the Seebeck coefficient, also decreased.

3.4. Test Fabrication of a Thermocouple

We tested the above findings by fabricating a thermocouple, which comprised p-type and n-type thermoelectric elements. These elements were fabricated selectively by controlling the laser irradiation conditions. The p-type element was formed at a writing speed of 500 μm/s and a laser fluence of 0.059 J/cm^2. The Seebeck coefficient of these p-type the micropatterns was expected to be 9.7 × 10^2 μV/K, as shown in Table 1. The n-type element was formed at a writing speed of 4000 μm/s and the laser fluence of 0.047 J/cm^2. The Seebeck coefficient of these n-type micropatterns was expected to be −19 μV/K, as listed in Table 3. All the elements were patterned with raster scanning of the laser focal point. The raster pitch was 10 μm, as discussed in the previous section. Figure 7a shows a photograph of the thermocouple. No damage to the micropatterns was apparent. The open-circuit voltage of the thermocouple was measured while a temperature difference between the sensing part and electrical contacts was generated. Figure 7b plots the relationship between the open-circuit voltage and the temperature difference. The voltage increased nearly linearly as the temperature difference increase. The difference from perfect linearity is possibly caused by the temperature dependence of Seebeck coefficient. A voltage of 51.2 mV was generated at the temperature difference of 89 K, which was smaller than the value of 88 mV estimated using the Seebeck coefficients of each element. The difference may be caused by the thermal history's effect on the micropatterns while the connected region of p- and n-type thermoelectric elements in the sensing tip was being fabricated at the two different laser-writing conditions.

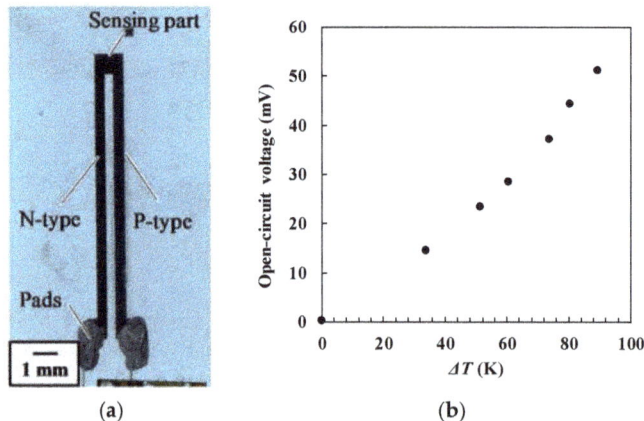

(a)　　　　　　　　　　　　　(b)

Figure 7. (a) Photograph of a sample thermocouple fabricated by selective micropatterning, (b) relationship between the open-circuit voltage and the temperature difference between sensing hot and cold.

4. Discussion

We investigated effects of heat accumulation on the fabrication of micropatterns using femtosecond laser reductive sintering of the CuO/NiO NPs. Even though the total incident energy for micropatterning was constant, variations in the laser fluence and writing speed affect the content of Cu, Ni, their alloys, and oxides. In the reductive sintering of CuO/NiO NPs, first, EG is dehydrated at 433–473 K and acetaldehyde is subsequently generated [15]. Then, acetaldehyde reduces NPs mixed with CuO and NiO into Cu and Ni NPs under sufficient energy from laser irradiation [8]. After that, Cu and Ni NPs form the Cu-Ni alloy [16]. When an excess of laser energy is applied to the mixed CuO/NiO NPs, the reduced Cu and Ni become reoxidized. Temperatures above 673 K lead Cu to form CuO [8,17]. In this case, composite metal oxide micropatterns are formed. If insufficient laser energy is applied to the mixed CuO/NiO NPs, some of the original CuO and NiO material is included in the micropatterns.

Even if the incident total energy was constant, the laser fluence affected the generation of the composite materials in our tests. The maximum temperature achieved decreased with decreasing laser fluence. The heating duration also increased with slower writing speeds. As a result, the fabricated micropatterns were easily reoxidized, as shown in Figure 5.

When the laser-writing speed and the incident total energy were constant, the use of single or double exposure affected the composites in the micropatterns. The proportion of metal oxides was higher with double exposure than that with single writing as shown in Figure 6. The maximum temperature achieved with the single writing at high laser fluence was higher than that with double exposure at low laser fluence. In addition, the total laser irradiation duration with the single writing was shorter than that at with the double exposure. These factors limit the diffusion of oxygen in the air to the micropatterns with the single-writing process.

Finally, the test fabrication of a thermocouple suggests the importance of the micropattern's thermal history. When the two micropatterns were selectively formed using two different laser irradiation conditions, each micropattern was differently thermally affected by the laser irradiation near the boundary of the two micropatterns. The composition and thermoelectric properties of the micropatterns may be controlled precisely by considering the thermal effects in the area surrounding the region of laser irradiation.

In this article, we experimentally and qualitatively investigated the patterning properties. However, the quantitative discussion is important. In the future, we will consider the phase transition using computational simulation of the temperatures.

5. Conclusions

We investigated the properties of micropatterns fabricated by the femtosecond laser reductive sintering of CuO/NiO mixed NPs. We found that heat accumulation during the laser irradiation process and the thermal history of the material significantly affected the composition and thermoelectric properties of the micropatterns fabricated by reductive sintering. Our tests show that selective fabrication of p-type and n-type thermoelectric micropatterns is possible with laser sintering of metal oxide NPs.

Author Contributions: K.N. performed the experiments; K.N. and M.M. analyzed the data; S.H. contributed analysis tools; M.M. wrote the paper.

Acknowledgments: This study was supported in part by the Nano-Technology Platform Program (Micro-Nano Fabrication), the Leading Initiative for Excellent Young Researchers (LEADER) of the Ministry of Education, Culture, Sports, Science and Technology, Japan (MEXT), the 10th "Shiseido Female Researcher Science Grant", and JSPS KAKENHI Grant number JP16H06064.

Conflicts of Interest: The authors have no conflict of interest to declare.

References

1. Tang, Y.; Loh, H.T.; Wong, Y.S.; Fuh, J.Y.H.; Lu, L.; Wang, X. Direct laser sintering of a copper-based alloy for creating three-dimensional metal parts. *J. Mater. Process. Technol.* **2003**, *140*, 368–372. [CrossRef]
2. Zhang, B.; Feinech, N.E.; Liao, H.L.; Coddet, C. Microstructure and magnetic properties of Fe–Ni alloy fabricated by selective laser melting Fe/Ni mixed powders. *J. Mater. Sci. Technol.* **2013**, *29*, 757–760. [CrossRef]
3. Sato, Y.; Tsukamoto, M.; Yamashita, Y. Surface morphology of Ti-6Al-4V plate fabricated by vacuum selective laser melting. *Appl. Phys. B* **2015**, *119*, 545–549. [CrossRef]
4. Dai, N.; Zhang, L.-C.; Zhang, J.; Zhang, X.; Ni, Q.; Chen, Y.; Wu, M.; Chao, C. Distinction in corrosion resistance of selective laser melted Ti-6Al-4V alloy on different planes. *Corros. Sci.* **2016**, *111*, 703–710. [CrossRef]
5. Theodorakos, I.; Zacharatos, F.; Geremia, R.; Karnakis, D.; Zergioti, I. Selective laser sintering of Ag nanoparticles ink for applications in flexible electronics. *Appl. Surf. Sci.* **2015**, *336*, 157–162. [CrossRef]
6. Watanabe, A. Laser sintering of metal nanoparticle film. *J. Photopolym. Sci. Technol.* **2013**, *26*, 199–205. [CrossRef]
7. Kwon, J.; Cho, H.; Eom, H.; Lee, H.; Suh, Y.D.; Moon, H.; Shin, J.; Hong, S.; Ko, S.H. Low-temperature oxidation-free selective laser sintering of Cu nanoparticle paste on a polymer substrate for the flexible touch panel applications. *Appl. Mater. Interfaces* **2016**, *8*, 11575–11582. [CrossRef] [PubMed]
8. Kang, B.; Han, S.; Kim, H.J.; Ko, S.; Yang, M. One-step fabrication of copper electrode by laser-induced direct local reduction and agglomeration of copper oxide nanoparticle. *J. Phys. Chem. C* **2011**, *115*, 23664–23670. [CrossRef]

9. Lee, H.; Yang, M. Effect of solvent and PVP on electrode conductivity in laser-induced reduction process. *Appl. Phys. A* **2015**, *119*, 317–323. [CrossRef]

10. Lee, D.; Paeng, D.; Park, H.K.; Grigoropoulos, C.P. Vacuum-free, maskless patterning of Ni electrodes by laser reductive sintering of NiO nanoparticle ink and its application to transparent conductors. *ACS Nano* **2014**, *8*, 9807–9814. [CrossRef] [PubMed]

11. Paeng, D.; Lee, D.; Yeo, J.; Yoo, J.H.; Allen, F.I.; Kim, I.; So, H.; Park, H.K.; Minor, A.M.; Grigoropoulos, C.P. Laser-induced reductive sintering of nickel oxide nanoparticles under ambient conditions. *J. Phys. Chem. C* **2015**, *119*, 6363–6372. [CrossRef]

12. Mizoshiri, M.; Ito, Y.; Sakurai, J.; Hata, S. Direct fabrication of Cu/Cu_2O composite micro-temperature sensor using femtosecond laser reduction patterning. *Jpn. J. Appl. Phys.* **2016**, *55*, 06GP05. [CrossRef]

13. Tamura, K.; Mizoshiri, M.; Sakurai, J.; Hata, S. Ni-based composite microstructures fabricated by femtosecond laser reductive sintering of NiO/Cr mixed nanoparticles. *Jpn. J. Appl. Phys.* **2017**, *56*, 06GN08. [CrossRef]

14. Mizoshiri, M.; Hata, S. Selective fabrication of p-type and n-type thermoelectric micropatterns by the reduction of CuO/NiO mixed nanoparticles using femtosecond laser pulses. *Appl. Phys. A* **2018**, *124*, 64. [CrossRef]

15. Fievet, F.; Lagier, J.P.; Blin, B.; Beaudoin, B.; Figlarz, M. Homogeneous and heterogeneous nucleations in the polyol process for the preparation of micron and submicron size metal particles. *Solid State Ion.* **2011**, *32–33*, 198–205. [CrossRef]

16. Wada, N.; Kankawa, Y.; Kaneko, Y. Comparison between Cu-Ni alloy and Cu, Ni powder mixture by metal injection molding. *J. Jpn Soc. Powder Metall.* **1997**, *44*, 604–611. [CrossRef]

17. Kevin, M.; Ong, W.L.; Ho, G.W. Formation of hybrid structures: Copper oxide nanocrystals templated on ultralong copper nanowires for open network sensing at room temperature. *Nanotechnology* **2011**, *22*, 235701. [CrossRef] [PubMed]

micromachines

MDPI

Article

Design of Substrate Stretchability Using Origami-Like Folding Deformation for Flexible Thermoelectric Generator

Kana Fukuie, Yoshitaka Iwata [ID] and Eiji Iwase *[ID]

Department of Applied Mechanics, Waseda University, 3-4-1 Okubo, Shinjuku-ku, Tokyo 169-8555, Japan; fukuie@iwaselab.amech.waseda.ac.jp (K.F.); iwata5116@ruri.waseda.jp (Y.I.)
* Correspondence: iwase@waseda.jp; Tel.: +81-3-5286-2741

Received: 11 May 2018; Accepted: 20 June 2018; Published: 22 June 2018

Abstract: A stretchable thermoelectric (TE) generator was developed by using rigid BiTe-based TE elements and a non-stretchable substrate with origami-like folding deformation. Our stretchable TE generator contains flat sections, on which the rigid TE elements are arranged, and folded sections, which produce and guarantee the stretchability of a device. First, a simple stretchable device with a single pair of p-type and n-type BiTe-based TE elements was designed and fabricated. The TE elements were sandwiched between two folded polyimide-copper substrates. The length of the wiring between the flat sections changed from 1.0 mm in the folded state to 1.8 mm in the deployed state. It was also confirmed that the single-pair device could generate power in both the folded and deployed states. After this, a stretchable TE generator with eight pairs of p-type and n-type BiTe-based TE elements connected in series was created. The stretchable TE generator was capable of withstanding a stretching deformation of 20% and could also produce an output voltage in both the folded and deployed states.

Keywords: stretchability; thermoelectric generator; flexible device; origami

---

1. Introduction

In this paper, a thermoelectric (TE) generator with bendability and stretchability is proposed in order to allow the TE generator to be attached to a non-flat surface. Conventionally, a TE generator that is used for recycling waste heat is composed of rigid TE elements and a ceramic plate. As a result, these conventional TE generators lack flexible properties, such as bendability and stretchability [1,2]. Therefore, its use is limited to a flat heat source. However, a flexible TE generator is required as there is frequently a need to attach generators to a heat source with a non-flat surface, such as piping or the human body. Recently, TE generators with bendability [3–12] and/or stretchability [13] were developed by using a deformable TE element (e.g., a carbon nanotube (CNT)-polystyrene (PS) composite) or a rubber-based stretchable substrate (e.g., polydimethylsiloxane (PDMS)). However, deformable TE conversion materials based on polymer composites unfortunately have inferior TE conversion efficiency compared to the rigid BiTe-based TE conversion materials, while the stretchable substrates using rubber have a narrow usable temperature range of 213–423 K [14].

Therefore, it is conceivable to realize a stretchable TE generator using a rigid BiTe-based TE conversion material with high TE conversion efficiency and origami-like folding deformation, which consists of a bendable but non-stretchable polymer substrate with a wide usable temperature range. Recently, the kirigami or origami-like structure has been used for stretchable devices [15–19]. Expansion-contraction deformation of the entire device is possible by connecting the rigid TE elements with foldable metal wiring. This is a feasible solution because expansion and contraction by folding does not require the substrate material to have stretchability. Thus, it is possible to select a material by considering its heat resistance and thermal characteristics. In addition, since folding occurs via local bending deformation, a flexible TE generator that is capable of deformation through expansion and contraction as well as bending can be realized by using a rigid TE element and non-stretchable substrate.

2. Experimental Section

2.1. Fabrication

The shape of the stretchable TE generator, which is proposed in this research, is shown in Figure 1, which demonstrates that it is capable of folding deformation. A polyimide-copper (Cu) film was used as a bendable but non-stretchable substrate with a wide usable temperature range. It is well known that polyimide has a wide usable temperature range of 4–673 K [20]. The stretchability of our TE generator is realized by connecting the rigid TE elements with the folded wiring parts and by reducing the distance between the TE elements. In this study, p-type and n-type TE elements were alternately arranged and were electrically connected in order to create a π structure.

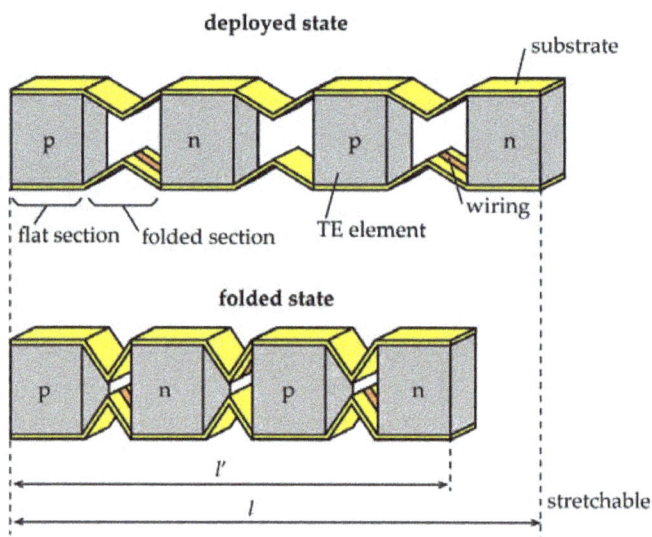

Figure 1. Stretchable substrate using origami-like folding deformation.

The fabrication process is shown in Figure 2. Using a cutting plotter (CE 6000-40, GRAPHTEC, Kanagawa, Japan), a polyimide-Cu substrate (Panasonic Corporation, Osaka, Japan. R-F 786 W; polyimide: 12.5 µm; Cu: 9 µm) was cut as shown in Figure 2a. The lower and upper substrates are shaped as squares of 16 × 16 mm with and without a pad for a lead wire, respectively, with each of these squares having nine 2-mm-long square holes. The cut polyimide-Cu substrate was fixed to a glass substrate using polyimide tape. A photoresist was spin-coated on the Cu layer and patterned to a wire shape. After the Cu layer was wet-etched, the photoresist was removed. Following this, both of the substrates were folded into a shape consisting of mountains and valleys (Figure 2b), thereby fabricating a substrate with stretchability. Finally, TE elements (Toshima Seisakusho, Saitama, Japan; p-type: $Bi_{0.3}Sb_{1.7}Te_3$; n-type: Bi_2Te_3) were mounted onto flat sections of the folded substrate using the cream solder (SMX-H05, Sun Hayato Co., Ltd., Tokyo, Japan). The size of each TE element is 2 mm × 2 mm × 1 mm.

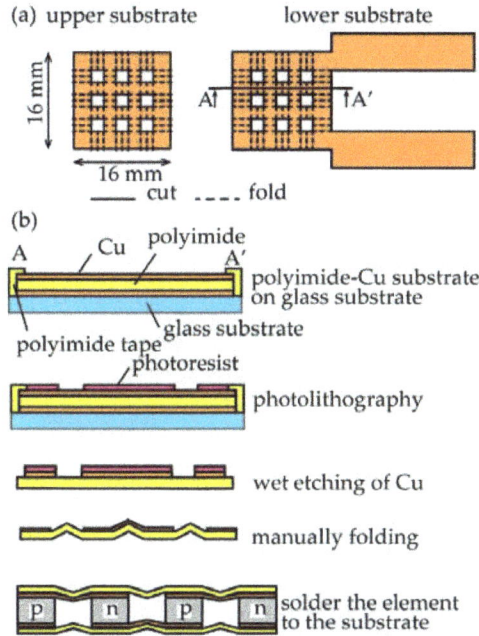

Figure 2. Design and fabrication process of the flexible thermoelectric (TE) generator: (**a**) Cutting and folding lines; and (**b**) fabrication process by photolithography.

2.2. Method

First, we aimed to confirm whether thermoelectricity could be generated using the stretchable substrate in its folded state. A pair of p-type and n-type TE elements were arranged in order to have a π-type structure, before a device in which the TE elements were connected by folded wiring was fabricated. The characteristics of the TE generator using folded wiring were evaluated by using a pair of TE generators. The electric power of the TE generator with respect to the temperature produced by the heat source before and after folding the pair of TE elements was measured. The open-circuit voltage V_{TEG} (V) of the TE generator is a value determined from the temperature difference that the TE generator is subjected to and the Seebeck coefficient of the device. Therefore, the Seebeck coefficient of the device can be obtained from the open-circuit voltage V_{TEG} (V) of the TE generator. Furthermore, the TE generator has an internal resistance R_{TEG}. Connecting the TE generator to the load causes a voltage drop to occur, whereupon the electric power changes according to the current flowing in the circuit. Therefore, the relationship between the voltage and the electric current with respect to the current and the internal resistance can be determined. The electric power that can be taken out of the TE generator was measured by placing a source meter in the load resistance part. First, a temperature difference was applied between the top and bottom surfaces of the TE generator. This was achieved by placing the TE generator on the ceramic heater and warming the bottom surface. A metal block of 20 mm × 21 mm × 30 mm was placed on the top surface to dissipate heat. The output of the TE generator was connected to a source meter (2614B, Keithley, Instruments, Cleveland, OH, USA) and voltage was applied, before the value of the current at that time was measured. The temperature of the ceramic heater was adjusted from 313 K to 393 K in increments of 20 K. The room temperature at the time of measurement was 297.8 K.

It is conceivable that the conditions of contact with the heat source might be different between a single pair and eight pairs of TE generators. Thus, the output voltage of the device with respect to the temperature provided by the heat source to the eight-pair TE generator before and after folding was also measured. A schematic diagram of the measurement setup is shown in Figure 3. A micro-ceramic heater was used to generate heat by supplying a voltage to the heater (MS-2, Sakaguchi E.H Voc Corp., Tokyo, Japan) using an AC/AC converter (Yamabishidenki Co., Ltd., Tokushima, Japan. bolt slider). The temperature of the micro-ceramic heater was measured by attaching a thermocouple to it. Feedback control using the temperature controller (E5CC-RW0AUM-000, Omuron, Kyoto, Japan) was performed to regulate the heat generation temperature of the micro-ceramic heater. Subsequently, the fabricated TE generator was placed on the micro-ceramic heater, while a cylindrical metal block with a diameter of 65 mm and a height of 28 mm was placed on the device in order to allow the top surface of the device to cool down. A digital multimeter (model 2000, Keithley Instruments, Cleveland, OH, USA) was used to measure the output voltage of the device. The temperature of the micro-ceramic heater was increased from 313 K to 393 K in increments of 20 K, with the output voltage of the device being measured during this period of time. The room temperature at the time of measurement was 297.7 K.

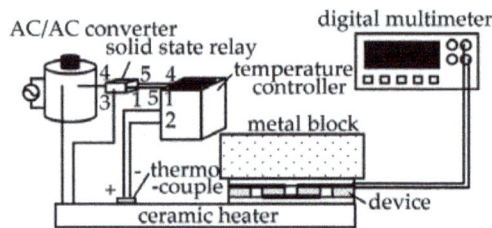

Figure 3. Experimental setup for measuring the output voltage of the flexible TE generator.

3. Result and Discussion

First, the stretchability of the fabricated devices, which are shown in Figure 4, was evaluated. In the deployed and folded states, the separation between the pair of TE generators was 1.8 mm and 1.0 mm, respectively (Figure 4a,b). Figure 4c shows the deployed device after mounting eight pairs of TE elements. The total length of the device, excluding the padding, was 15 mm. On the other hand, Figure 4d shows the entire device in the folded state with a total length of 12 mm. This result indicates that a TE generator that is capable of a stretching deformation of 20% with respect to the deployed devices was realized.

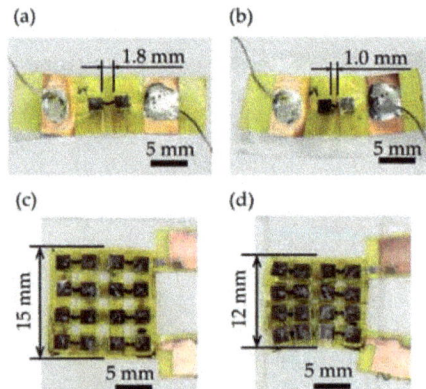

Figure 4. Photographic images of the fabricated devices. A single pair of TE generators in (**a**) the deployed state and (**b**) the folded state. Eight pairs of TE generators in (**c**) the deployed state and (**d**) the folded state.

Next, the basic characteristics of the TE generator containing a single pair of TE elements were measured. Figure 5a,b show the relationships between the current value and input voltage of the deployed and folded states, respectively. The TE generator has good linearity in both the deployed and folded states. The relationship between the input voltage and current of the TE generator can be expressed as:

$$V_L = -R_{TEG}I + V_{TEG} \qquad (1)$$

where V_{TEG} (V) is the open-circuit voltage of the TE generator, V_L (V) is the input voltage of the source meter, R_{TEG} (Ω) is the internal resistance of the TE generator and I (A) is the current measured by the source meter. Therefore, the internal resistance R_{TEG} of the TE generator with a single pair of TE elements was calculated from the gradients of the fitting lines in Figure 5a,b. The open-circuit voltage with respect to the temperature can be obtained. Figure 5c shows a plot of the internal resistance of the single pair of TE elements obtained from the slope of Figure 5a,b. Figure 5d plots the open-circuit voltage of the TE generator with the single pair of TE elements with respect to the temperature of the ceramic heater from Figure 5a,b. Since it is an electromotive force, the absolute values were taken and plotted. The output voltage V_{TEG} is V_L when $I = 0$. The fitting lines of the deployed and folded states have the regression coefficients $R^2 = 0.993$ and $R^2 = 0.999$, respectively. When the ceramic heater temperature was 393 K, the deployed device received 10.3-mV and the folded device received 8.90-mV open circuit voltage. When the temperature of the ceramic heater increased, a temperature difference was generated between the top and the bottom surfaces of the device, resulting in an increase in the open-circuit voltage. The Seebeck coefficient of the device, which is the regression coefficient of the filling lines in Figure 5d, was 0.106 ± 0.005 mV/K and 0.090 ± 0.001 mV/K for the deployed and folded states, respectively. The Seebeck coefficients measured for both states are close to each other, but statistically different. The reason for the difference in the Seebeck coefficients might be that the contact thermal resistance in the folded device is higher. The relationship between the electric power and the temperature of the ceramic heater is shown in Figure 5e. As the temperature difference increases, the output also increases. When the temperature of the ceramic heater was 393 K, this resulted in an output power of 56.8 μW for the deployed device and 40.7 μW for the folded device. It was possible to generate power with the TE generator using folded wiring. In Figure 5, the device characteristics were evaluated by using the heating temperature, because the contact thermal resistance change should be included in the device characteristics in our experiments. In general, the temperature difference within the TE generator is proportional to the supplied temperature difference [21], which is the temperature difference between the hotplate and room temperature (Figure 5f). The difference between the gradients of the fitting lines in Figure 5c,d originating from the change in the contact thermal resistance was considered. In this experiment, since a metal block was placed on the device, the pressure, which should be the reduced difference in contact thermal resistance, was applied to the device. Therefore, it is conceivable that the thermal contact conditions between the deployed state and folded state differed according to folding deformation.

The open-circuit voltage was also measured for the TE generator with eight pairs of TE elements. This is because the area, in which the heat source is installed and the contact thermal conductivity is affected, can be varied by using either a single pair or eight pairs of TE elements. The result of the output voltage with respect to the temperature of the ceramic heater is shown in Figure 6. The result confirmed that the fitting line for each device in the deployed and folded states has good linearity. In both cases, the output voltage also tends to increase in proportion to the temperature input to the lower substrate. Consequently, the performance of the device as a TE generator is effective. However, the output voltage of the folded state is lower than that of the deployed device. The Seebeck coefficient of the device was 0.61 ± 0.03 mV/K and 0.49 ± 0.01 mV/K for the deployed and folded states, respectively. The changing ratio of the Seebeck coefficient of the device for the eight-pair device in Figure 6 is larger than that for the single-pair device in Figure 5d. We believe that this is because the contact thermal resistance can easily increase due to the difficulty of maintaining uniform folding in a device with a large area. It is difficult to transfer heat due to the creation of a gap between the substrate and the heater or the metal block. The above results confirm that it is possible to realize a TE generator using a substrate with a folded structure.

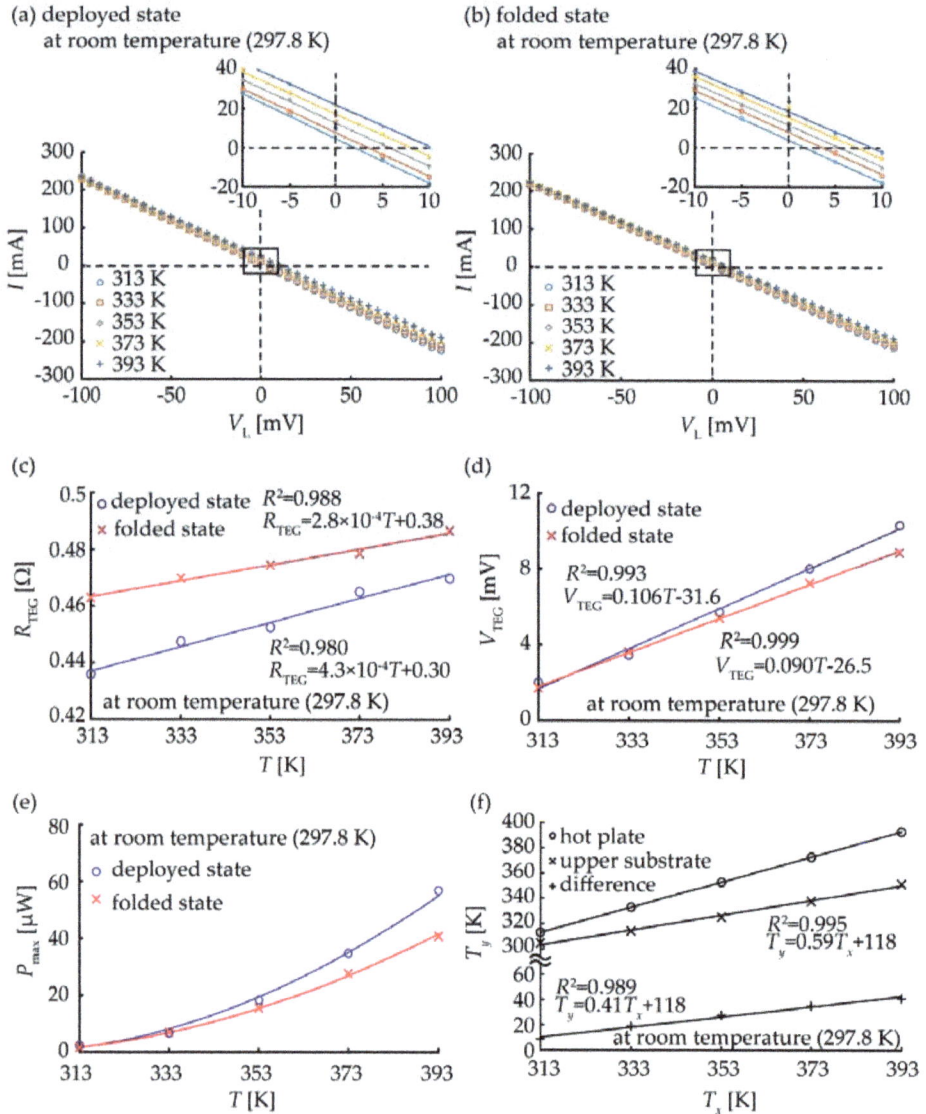

Figure 5. Measurement results of the characteristics of the devices in the case of a single pair of p-type and n-type TE conversion elements: Relationship of the output current to the input voltage of the source meter for the (**a**) deployed and (**b**) folded states; (**c**) Relationship between the internal resistance and the temperature of each of the deployed and folded states; (**d**) Relationship between open circuit voltage and temperature; (**e**) Relationship between maximum output power and temperature; and (**f**) Temperature difference between the top and bottom surfaces of a single pair of p-type and n-type TE elements in the deployed state.

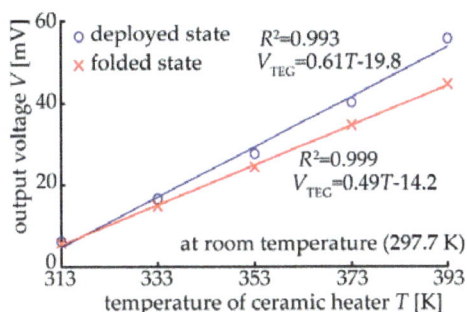

Figure 6. Relationship between the temperature of ceramic heater and output voltage.

4. Conclusions

In this study, we realized a stretchable TE generator with a deformation ratio of 20%. The rigid BiTe-based TE elements are connected with folded wiring. A device consisting of a single pair of TE elements was fabricated and the TE characteristics of this device were evaluated. For the single pair of elements, it was confirmed that electric power is generated because of the difference in temperature between the upper and lower surfaces of the device. After this, a TE generator with a larger area was demonstrated by electrically connecting a pair of devices in series using folded wiring. The measurements showed that the output voltage of the TE generator that is capable of expansion and deformation tends to rise in proportion to the input temperature in the case of both the deployed and folded devices. The Seebeck coefficient of the devices was different between the deployed and folded state. This is because the contact thermal resistance becomes high due to folding deformation. In this research, a stretchable TE generator using a non-stretchable substrate was developed. This means that a stretchable TE generator with few constraints can be fabricated from substrate materials. Therefore, since materials with good thermal conductivity could also be used, it is possible to realize a highly efficient flexible TE generator by using our method.

Author Contributions: K.F., Y.I., and E.I. conceived and designed the experiments; K.F. performed the experiments, analyzed the data, and wrote the paper; and E.I. supervised the research.

Funding: This research was partially funded by JST CREST grant number JPMJCR16Q5.

Acknowledgments: The authors thank the MEXT Nanotechnology Platform Support Project of Waseda University.

Conflicts of Interest: The authors declare no conflicts of interest.

References

1. Rowe, D.M.; Min, G. Design theory of thermoelectric modules for electrical power generation. *IEE Proc. Sci. Meas. Technol.* **1996**, *143*, 351–356. [CrossRef]
2. Snyder, G.J.; Toberer, E.S. Complex thermoelectric materials. *Nat. Mater.* **2008**, *7*, 105–114. [CrossRef] [PubMed]
3. Suemori, K.; Hoshino, S.; Kamata, T. Flexible and lightweight thermoelectric generators composed of carbon nanotube-polystyrene composites printed on film substrate. *Appl. Phys. Lett.* **2013**, *103*, 153902. [CrossRef]
4. Kim, S.J.; We, J.H.; Cho, B.J. A wearable thermoelectric generator fabricated on a glass fabric. *Energy Environ. Sci.* **2014**, *7*, 1959–1965. [CrossRef]
5. Glatz, W.; Schwyter, E.; Durrer, L.; Hierold, C. Bi₂Te₃-based flexible micro thermoelectric generator with optimized design. *J. Microelectromech. Syst.* **2009**, *18*, 763–772. [CrossRef]
6. Kim, M.K.; Kim, M.S.; Jo, S.E.; Kim, Y.J. Flexible thermoelectric generator for human body heat energy harvesting. *Electron. Lett.* **2012**, *48*, 1015–1017. [CrossRef]
7. Francioso, L.; De Pascali, C.; Sglavo, V.; Grazioli, A.; Masieri, M.; Siciliano, P. Modelling, fabrication and experimental testing of an heat sink free wearable thermoelectric generator. *Energy Convers. Manag.* **2017**, *145*, 204–213. [CrossRef]

8.   Yang, C.; Souchay, D.; Kneiß, M.; Bogner, M.; Wei, H.M.; Lorenz, M.; Oeckler, O.; Benstetter, G.; Fu, Y.Q.; Grundmann, M. Transparent flexible thermoelectric material based on non-toxic earth-abundant p-type copper iodide thin film. *Nat. Commun.* **2017**, *8*, 4–10. [CrossRef] [PubMed]

9.   Shi, Y.; Wang, Y.; Mei, D.; Feng, B.; Chen, Z. Design and Fabrication of wearable thermoelectric generator device for heat harvesting. *IEEE Robot. Autom. Lett.* **2018**, *3*, 373–378. [CrossRef]

10. Nishino, T.; Suzuki, T. Flexible thermoelectric generator with efficient vertical to lateral heat path films. *J. Micromech. Microeng.* **2017**, *27*, 035011. [CrossRef]

11. Trung, N.H.; Van Toan, N.; Ono, T. Fabrication of π-type flexible thermoelectric generators using an electrochemical deposition method for thermal energy harvesting applications at room temperature. *J. Micromech. Microeng.* **2017**, *27*, 125006. [CrossRef]

12. Wang, X.; Meng, F.; Wang, T.; Li, C.; Tang, H.; Gao, Z.; Li, S.; Jiang, F.; Xu, J. High performance of PEDOT:PSS/SiC-NWs hybrid thermoelectric thin film for energy harvesting. *J. Alloys Compd.* **2018**, *734*, 121–129. [CrossRef]

13. Jeong, S.H.; Cruz, F.J.; Chen, S.; Gravier, L.; Liu, J.; Wu, Z.; Hjort, K.; Zhang, S.L.; Zhang, Z.B. Stretchable thermoelectric generators metallized with liquid alloy. *Appl. Mater. Interfaces* **2017**, *9*, 15791–15797. [CrossRef] [PubMed]

14. Characteristic Properties of Silicone Rubber Compounds. Available online: https://www.shinetsusilicone-global.com/catalog/pdf/rubber_e.pdf (accessed on 31 May 2018).

15. Rojas, J.P.; Conchouso, D.; Arevalo, A.; Singh, D.; Foulds, I.G.; Hussain, M.M. Paper-based origami flexible and foldable thermoelectric nanogenerator. *Nano Energy* **2017**, *31*, 296–301. [CrossRef]

16. Song, Z.; Wang, X.; Lv, C.; An, Y.; Liang, M.; Ma, T.; He, D.; Zheng, Y.-J.; Huang, S.-Q.; Yu, H.; et al. Kirigami-based stretchable lithium-ion batteries. *Sci. Rep.* **2015**, *5*, 10988. [CrossRef] [PubMed]

17. Hwang, D.-G.; Bartlett, M.D. Tunable mechanical metamaterials through hybrid kirigami structures. *Sci. Rep.* **2018**, *8*, 3378. [CrossRef] [PubMed]

18. Lamoureux, A.; Lee, K.; Shlian, M.; Gorrest, S.R.; Shtein, M. Dynamic kirigami structures for integrated solar tracking. *Nat. Commun.* **2015**, *6*, 8092. [CrossRef] [PubMed]

19. Rojas, J.P.; Singh, D.; Arevalo, A.; Foulds, I.G.; Hussain, M.M. Stretchable helical architecture inorganic-organic hetero thermoelectric generator. *Nano Energy* **2016**, *30*, 691–699. [CrossRef]

20. DEC-Kapton-Summary-of-Properties.pdf. Available online: http://www.dupont.com/content/dam/dupont/products-and-services/membranes-and-films/polyimde-films/documents/DEC-Kapton-summary-of-properties.pdf (accessed on 31 May 2018).

21. Baranowski, L.L.; Snyder, G.J.; Toberer, E.S. Effective thermal conductivity in thermoelectric materials. *J. Appl. Phys.* **2013**, *113*, 204904. [CrossRef]

micromachines

MDPI

Article

Possibility of Controlling Self-Organized Patterns with Totalistic Cellular Automata Consisting of Both Rules like Game of Life and Rules Producing Turing Patterns

Takeshi Ishida [iD]

Department of Ocean Mechanical Engineering, National Fisheries University, Shimonoseki 759-6595, Japan; ishida07@ecoinfo.jp; Tel.: +81-832-86-5111

Received: 20 April 2018; Accepted: 28 June 2018; Published: 3 July 2018

Abstract: The basic rules of self-organization using a totalistic cellular automaton (CA) were investigated, for which the cell state was determined by summing the states of neighboring cells, like in Conway's Game of Life. This study used a short-range and long-range summation of the cell states around the focal cell. These resemble reaction-diffusion (RD) equations, in which self-organizing behavior emerges from interactions between an activating factor and an inhibiting factor. In addition, Game-of-Life-type rules, in which a cell cannot survive when adjoined by too many or too few living cells, were applied. Our model was able to mimic patterns characteristic of biological cells, including movement, growth, and reproduction. This result suggests the possibility of controlling self-organized patterns. Our model can also be applied to the control of engineering systems, such as multirobot swarms and self-assembling microrobots.

Keywords: cellular automata; Game of Life; reaction-diffusion system; self-organization; Turing pattern model; Young model

1. Introduction

Self-organization phenomena, in which global structures are produced from purely local interactions, are found in fields ranging from biology to human societies. If these emergent processes were to be controlled, various applications would be possible in engineering fields. For this purpose, the conditions under which they emerge must be elucidated. This is one of the goals of the study of complex systems. Such control methods may allow the automatic construction of machines. They could also be applied to the control of engineering systems, such as controlling robot swarms or self-assembling microrobots. In addition, the methods allow for the growth of artificial organs or the support of ecosystem conservation, as well as clarifying the emergence of the first life on ancient earth.

Mathematical modeling of self-organization phenomena has two main branches: the mathematical analysis of reaction-diffusion (RD) equations, and discrete modeling using cellular automata (CA).

The Turing pattern model is one type of RD model. This was introduced by Turing in 1952 [1], where he treated morphogenesis as the interaction between activating and inhibiting factors. Typically, this model achieves self-organization through the different diffusion coefficients for two morphogens, equivalent to an activating and an inhibiting factor. The general RD equations can be written as follows:

$$\frac{\partial u}{\partial t} = d_1 \nabla^2 u + f(u, v),$$

$$\frac{\partial v}{\partial t} = d_2 \nabla^2 v + g(u, v),$$

where u and v are the morphogen concentrations, functions f and g are the reaction kinetics, and d_1 and d_2 are the diffusion coefficients. Previous studies have considered a range of functions f and g, and models such as the linear model, the Gierer–Meinhardt model [2], and the Gray–Scott model [3] have been used to produce typical Turing patterns.

By contrast, CA models are discrete in both space and time. The state of the focal cell is determined by the states of the adjacent cells and the transition rules. The advantage of CA models is that they can describe systems that cannot be modeled using differential equations. Historically, various interesting CA patterns have been discovered.

In the 1950s, von Neumann introduced the mathematical study of self-reproduction phenomena, and proved, theoretically, that a self-replicating machine could be constructed on a two-dimensional (2D) grid using 29 cell states and transition rules [4]. Von Neumann's self-reproducing machine was too large to be implemented until powerful computers became available in the 1990s [5]. Codd [6] showed that the number of cell states could be reduced to eight, and proved the possibility of a self-reproducing machine. However, this model was too complex to be applied to biological processes. In the 1970s, the Game of Life (GoL), invented by Conway [7], became popular. It is a simple CA model that is defined by only four rules, yet produces complex dynamic patterns. Wolfram [8] investigated the one-dimensional (1D) CA model and identified four categories of pattern formation. One of these was the chaotic behavior group. Langton [9] developed a simple self-reproducing machine that did not require von Neumann's machine completeness. While having a very simple shape, Langton's machine's transition rules were complex and produced only specific circling shapes. Byl [10] simplified the Langton model, and Reggia [11] identified the simplest possible reproducing machine. Ishida [12] demonstrated a self-reproducing CA model that resembled a living cell, with DNA-like information carriers held inside a cell-like structure.

On the other hand, many studies have applied CA models to biological phenomena, such as the generation of seashell-like structures by Wolfram [8], the development by Young of an RD model in a CA setting that reproduced the patterns of animal coloration [13], the application of the Belousov–Zhabotinskii reaction CA model by Madore and Freeman [14], the catalytic modeling of proteins by Gerhardt and Schuster [15], the modeling of the immune system by De Boer and Hogeweg [16] and by Celada and Seiden [17], the tumor growth model of Moreira and Deutsch [18], the genetic disorder model of Moore and Hahn [19], the modeling of the hippocampus by Pytte et al. [20], and the dynamic modeling of cardiac conduction by Kaplan et al. [21]. Furthermore, Markus [22] demonstrated that a CA model could produce the same output as RD equations.

In addition, CA has two types of model. The first one is a totalistic CA, for which each cell state is determined by summing the states of the neighboring cells. Conway's GoL is one such CA. GoL produces the three typical patterns from the particular initial cell state configurations shown in Figure 1: "still", "moving", and "oscillating". Meanwhile, most of the initial cell state configurations reach extinction, i.e., all cell states become 0. However, it is impossible to deduce the transition rules to make specific shapes automatically.

In contrast with the totalistic CA, the second type of CA model counts all the neighbor cells' state patterns. Wolfram's 1D CA is one such CA, in which there are two states (0 and 1) of three inputs and one output. As there are $2^3 = 8$ inputs, and each input has two outputs (0 or 1), there are $2^8 = 256$ transition rules. Wolfram researched all these transition rules and divided the output patterns into four classes. Class 1 covers the steady states, Class 2 the periodical patterns, Class 3 the random patterns (edge of chaos), and Class 4 the chaotic patterns. Figure 2 shows one result of the Class 3 rules, which produce unique patterns like that of a certain type of seashell.

However, in a Wolfram-type model, increasing the inputs to five produces an explosive increase in the number of output combinations, to 2^{32}. It therefore becomes impossible to investigate all outputs. A similar situation occurs in the case of a 2D CA model, making it, again, impossible to investigate all possible patterns. Langton proposed the λ parameter [23] to elucidate the condition to give rise to the diversity of patterns—edge of chaos. However, this could not derive the rule to construct favorite patterns.

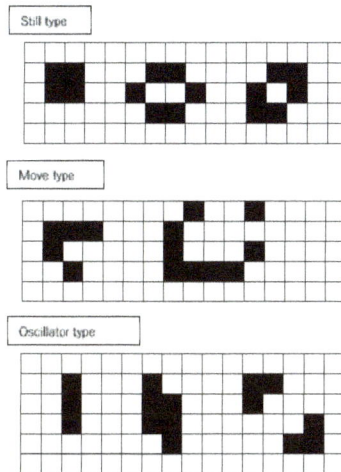

Figure 1. Patterns from Conway's Game of Life under four rules: (1) a cell with one or no living neighbors dies; (2) a cell with four or more living neighbors dies; (3) a cell with three living neighbors is born; and (4) a cell with two or three living neighbors survives. The typical patterns that emerge can be categorized into three groups: still, moving, or oscillation.

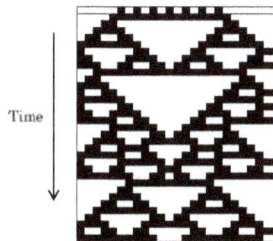

Figure 2. One of the patterns produced by Wolfram's 1D cellular automaton (CA) model. Rule 126: white cell indicates the 0 state, while the black cell indicates the 1 state. The upper line corresponds to the initial condition. Applying this rule, the time series pattern produces a range of triangular shapes, in this case resembling a certain type of seashell.

CA models reduce the computational load by removing the need to solve differential equations numerically. However, it is necessary to translate the partial differential equations of the RD equations into the transition rules of the CA model. In a study of up to present, no general method currently exists for doing this, with the exception of the ultra-discrete systems [24] used in certain soliton equations. Normally, the transition rules driving the emergent phenomena must be found by trial and error. Studies that attempted to derive the transition rules using genetic algorithms [11,25] demonstrated the difficulty of deriving generalized rules. The Ishida cell division model [12] is a hybrid of the Wolfram- and Conway-type CAs, in which part of the rules are totalistic transition rules, and the others are head-to-head-type transition rules, and these rules were discovered by trial and error.

The Young model [13] is one of the 2D totalistic models that bridges the RD equations and CA model; this model is used to produce Turing patterns. Some other examples to produce Turing patterns are below. Adamatzky [26] studied a binary-cell-state eight-cell neighborhood two-dimensional cellular automaton model with semitotalistic transitions rules. Dormann [27] also used a 2D outer-totalistic model with threes states to produce a Turing-like pattern. Tsai [28] analyzed a self-replicating mechanism in Turing patterns with a minimal autocatalytic monomer-dimer system.

Young's CA model uses a real number function to derive the distance effects, with two constant values within a grid cell: u_1 (positive) and u_2 (negative). The calculation begins by randomly distributing black cells on a rectangular grid. Then, for each cell at position R_0 in 2D fields, the next cell state of R_0, due to all nearby black cells at position R_i, are added up. R_i is assumed to be a black cell within radius r_2 from R_0 cell. The function v(r) is a continuous step function, as shown in Figure 3b. The activation area, indicated by u_1, has a radius of r_1 and the inhibition area, indicated by u_2, has a radius of r_2 ($r_2 > r_1$). At position R_0, when R_i is assumed to be a grid within r_2, function $v(|R_0 - R_i|)$ value is added up. Function $|R_0 - R_i|$ indicates the distance between R_0 and R_i. If $\sum_i v(|R_0 - R_i|) > 0$, the grid cell at point R_0 is marked as a black cell. If $\sum_i v(|R_0 - R_i|) < 0$, the grid cell becomes a white cell. If $\sum_i v(|R_0 - R_i|) = 0$, the grid cell does not change state [13]. Using these functions, Young described that a Turing pattern can be generated. Spot patterns or striped patterns can be created with relative changes between u_1 and u_2.

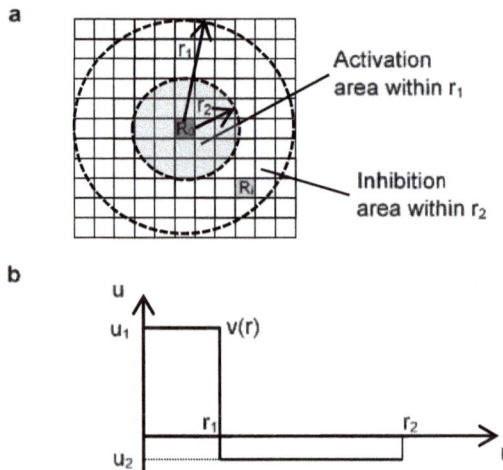

Figure 3. Outline of Young's model. (**a**) The activation area has a radius r_1 and the inhibition area has an outer radius r_2; (**b**) Function v(r) is a continuous step function representing the activation area and inhibition area.

In this Young model, let $u_1 = 1$, $u_2 = w$ (here $0 < w < 1$), and further, if the state of the cell is set to 0 (white) and 1 (black), this model can be arranged as follows. The state of cell i is expressed as $c_i(t)$ ($c_i(t) = [0, 1]$) at time t. The following state $c_i(t+1)$ at time $t+1$ is determined by the states of the neighboring cells. Here, N_1 is the sum of the states of the domain within S meshes of the focal cell. Similarly, N_2 is the sum of the states of the domain within t meshes of the focal cell, assuming that $S < t$.

$$N_1 = \sum_{i=1}^{S} c_i(t)$$

$$N_2 = \sum_{i=1}^{T} c_i(t)$$

Here, S (T) is the number of cells within s (t) meshes from focal cell. Figure 4 shows the schematic of the total sum of states N_1 and N_2. The next time state of the focal cell is determined by the following Expression (1):

$$\text{Cell state at the next time step} = \begin{cases} 1 : \text{if } N_1 > N_2 \times w \\ \text{Unchange} : \text{if } N_1 = N_2 \times w \\ 0 : \text{if } N_1 < N_2 \times w \end{cases} \qquad (1)$$

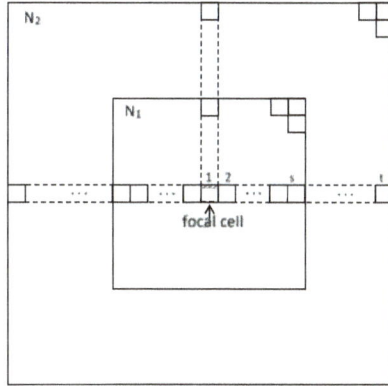

Figure 4. Schematic of the summing of states N_1 and N_2. Each grid cell has state 0 (white) or 1 (black). The inner area has a domain within s grids of the focal cell and the outer area a domain within t grids.

The essence of the Turing pattern model is that the short- and long-range spatial scales are each affected by separate factors [29], and pattern formation emerges from nonlinear interactions between the two factors. Turing used two chemicals with different diffusion coefficients to create these short- and long-range spatial effects. However, as long as there exists a difference between long- and short-range effects, other implementation methods can be applied. This model used two ranges of s mesh and t mesh to create a difference. It is therefore thought that this model is essentially the same as an RD model.

Figure 5 shows the results for the Expression (1) model using a square grid. As the initial conditions, 300 state 1 cells were set randomly in the calculation field 80 × 80. Turing-like patterns emerged as parameter w changed. When w was greater than 0.4, spot patterns were observed, whereas at $w = 0.3$, stripe patterns emerged. When w was 0.20 or lower, all cells in the field had a value of 1 (shown in black). Changing the parameters N_1 and N_2 merely changed the scale of the patterns.

On the other hand, the GoL is one of the 2D totalistic CAs to emerge patterns of diversity. In this model, the state of the focal cell is activated when the total number of surrounding states is within a certain range. Therefore, we considered the extended model of GoL. Whereas the GoL uses the states of the eight surrounding cells (the Moore neighborhood), an extended neighborhood is possible that considers two or more meshes from the focal cell like Young model. Such models are expected to generate a similar variety of patterns to the GoL.

In natural phenomena, short-distance influences play a stronger role than more distant influences. Under this circumstance (1) The influence decreases as the distance from the focal cell increases, which is $N_1 + N_2 \times w$ type model (w: weight parameter, $0 < w < 1$). However, reaction diffusion phenomena are created with long-distance influences that play a negative role than more near influences. (2) The influence reverses as the distance from the focal cell increases, which is $N_1 - N_2 \times w$ type model (w: weight parameter, $0 < w < 1$).

Initial condition

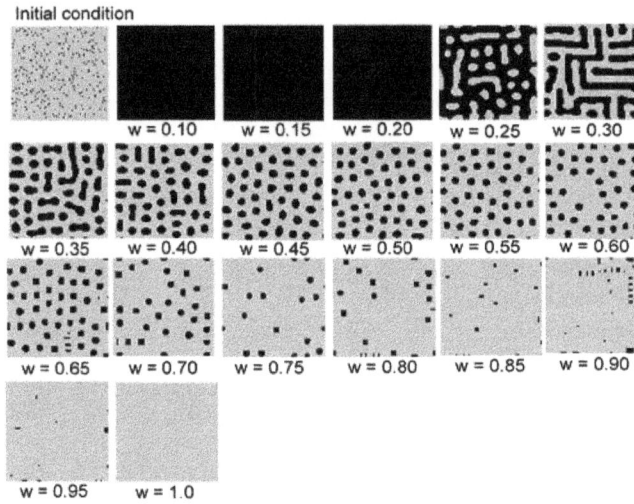

Figure 5. Results for the model of Expression (1) with $s = 5$ and $t = 9$ in a square grid. As the initial condition, 300 state 1 cells were set randomly in the calculation field. Turing-like patterns emerged as parameter w changed. When w was greater than 0.4, spot patterns were observed, whereas at $w = 0.3$, stripe patterns emerged. When w was 0.20 or lower, all cells in the field had a value of 1 (shown in black).

The first type of the model is expected to produce results similar to those of other totalistic CAs. Under specific conditions, GoL-like patterns will be found. The second model is expected to produce results similar to those of a Young model, which similarly takes account of the nearness of the activating and inhibiting factors.

This study adopted the second type of the model, and the emergence of a variety of patterns, such as the Turing pattern, were confirmed. In addition, by applying GoL-type rules to the second model type, more complex patterns are expected, such as self-reproducing and growth patterns. Our study model was able to produce not only a Turing-like pattern while using a simple transition expression, but also the model produces various patterns such as still, moving, and oscillation. Furthermore, this model can control these patterns with two parameters, which means the possibility of controlling self-organized patterns with a totalistic CA model.

2. Model

We investigated the results of an $N_1 - N_2 \times w$ type totalistic CA model in a 2D field that was implemented in Java. The GoL-like rules assumed in this model which limit the range of survival as state 1. To address this, by extending Expression (1), we introduced Expression (2). This expression was obtained by adding Rule 4 to Expression (1). When $N_2 = 0$, it is assumed that Rule 4 takes precedence.

$$\text{Cell state of the next time step} = \begin{cases} 0 : \text{if } N_1 > N_2 \times w_2 \ (Rule\ 4) \\ 1 : \text{if } N_1 > N_2 \times w_1 \ (Rule\ 3) \\ \text{Unchange} : \text{if } N_1 = N_2 \times w_1 (Rule\ 2) \\ 0 : \text{if } N_1 < N_2 \times w_1 (Rule\ 1) \end{cases} \quad (2)$$

$$N_1 = \sum_{i=1}^{S} c_i(t)$$

$$N_2 = \sum_{i=1}^{T} c_i(t)$$

Here, $1 > w_2 > w_1 > 0$, and N_1 is the sum of the states of the domain within s meshes of the focal cell. Similarly, N_2 is the sum of the states of the domain within t meshes of the focal cell, assuming that $s < t$. S (T) is the number of cells within s (t) meshes from focal cell. As with the Young model, we used "Unchange" as the equality in Expression (2) to prevent all cells becoming state 1 or 0 in the next time step when $N_1 = 0$ and $N_2 = 0$.

The model used 2D hexagonal grids (Figure 6), in which the transition rules were simple to apply. Although square grids are generally used in 2D CA modeling, we also used hexagonal grids for the reason that a hexagonal grid is isotropic as opposed to a square grid. The models were implemented under the following conditions:

- Calculation field: 80×80 cells in hexagonal grids
- Periodic boundary condition
- Initial conditions were assumed following two types of conditions:

  Condition (a): some state 1s were set randomly in the calculation field;
  Condition (b): some state 1s were set in the center of the calculation fields (Figure 7).

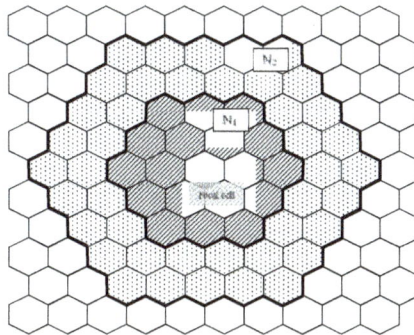

Figure 6. Hexagonal grid field. Here, N_1 is the domain within s meshes of the focal cell. Similarly, N_2 is the sum of the states of the domain within t meshes of the focal cell, assuming that $s < t$.

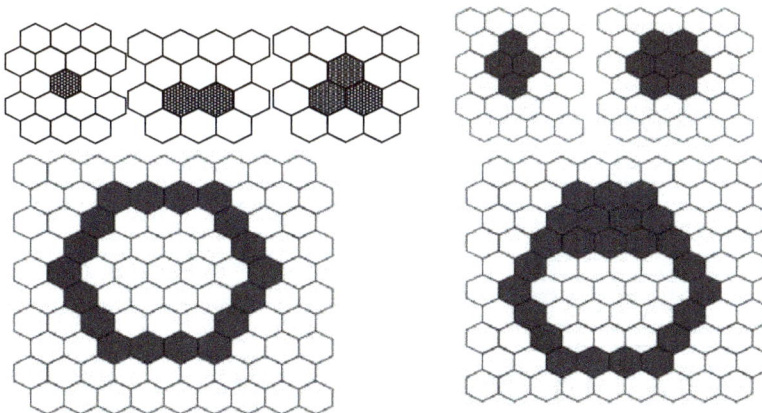

Figure 7. Initial conditions of hexagonal grids field. The black cell indicates the 1 state and the white cell indicates the 0 state. These configurations of state 1s were set in the center of the calculation fields.

3. Results

Figure 8 maps the model results against parameters w_1 and w_2. Initial condition is based on 300 randomly arranged state 1 cells. Different patterns were observed at different values of w_1 and w_2. As in the Young-type model, the parameter w_1 determined the pattern type, whether spotted or striped. As w_2 became smaller, pulsing and unsteady white spots were born within the black pattern. The vibration of these white spots became more intense as the value of w_2 was reduced. Figure 9 shows the results when varying the number of states 1 (black cell) to place them randomly in the initial state. The calculation conditions were $s = 5$, $t = 9$, $w_1 = 0.40$, and $w_2 = 0.80$ in a hexagonal grid. Based on the results of this figure, if there are more than a certain number of states 1, it is evident that a similar pattern formation develops. Furthermore, Figure 10 shows the transition of the ratio of states 1 and 2, and the ratio of the applied cells of Rule 4 at each time step. This result was calculated when w_2 was changed ($s = 5$, $t = 9$, and $w_1 = 0.35$). When w_2 was 1, Rule 4 was not applied, and the result was equivalent to the Young model. As w_2 was lowered, the application ratio of Rule 4 increased. At this time, white regions were generated in the black spots, and the instability increased and became pulsating. This phenomenon is attributed to Rule 4.

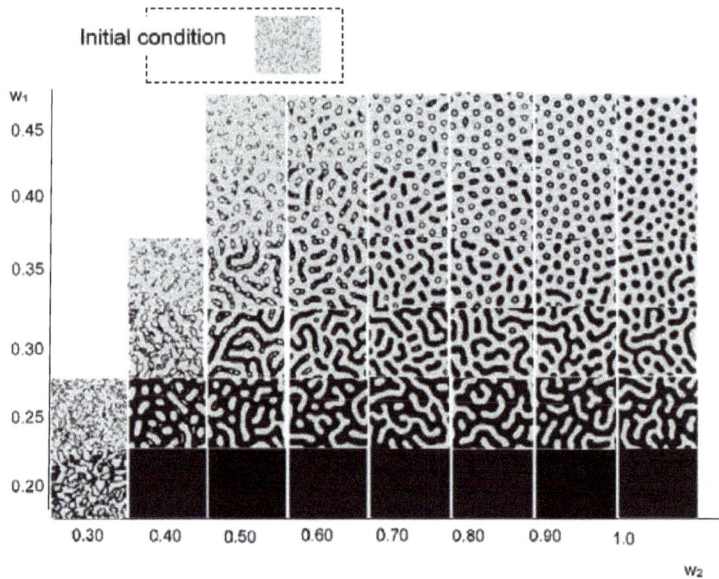

Figure 8. Results for the model with $s = 5$ and $t = 9$ in a hexagonal grid. The results image of each w_1 and w_2 value is based on 300 randomly arranged state 1 cells. Instability became stronger as w_2 decreased to w_1.

Figure 11 shows the effect of the initial condition which is centrally arranged on four black cells. The instability became more pronounced as w_2 decreased, and the central spotted pattern grew as an unstable wave in the calculation field. By contrast, when w_2 was relatively large, the central spot did not spread.

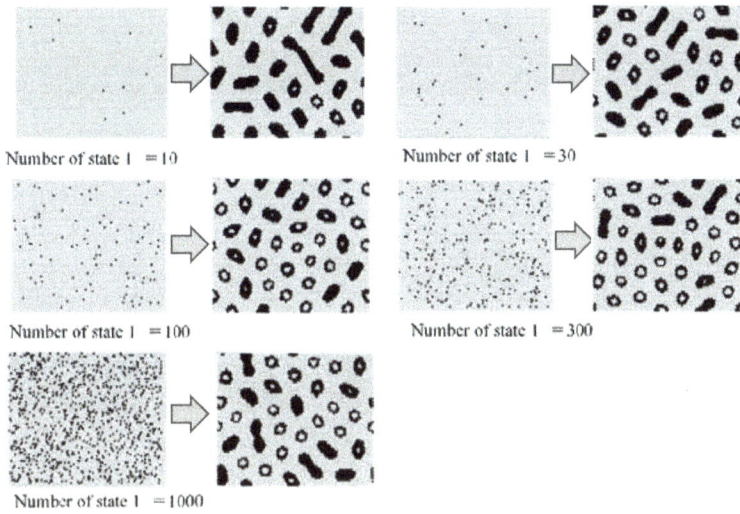

Figure 9. Results when varying the number of states 1 (black cell) to place them randomly in the initial states. The calculation conditions were $s = 5$, $t = 9$, $w_1 = 0.40$, and $w_2 = 0.80$ in a hexagonal grid. Based on this result, if there are more than a certain number of states 1, it is evident that a similar pattern formation develops.

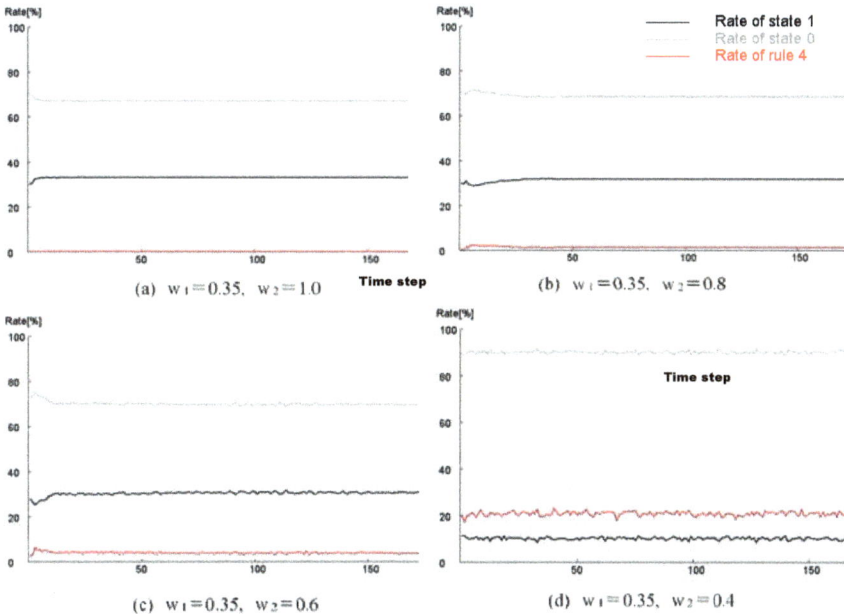

Figure 10. Transition ratio of states 1 and 2, and the ratio of the applied cells of Rule 4 at each time step. This result was calculated when w_2 was changed in $s = 5$, $t = 9$, and $w_1 = 0.35$. When w_2 was 1, Rule 4 was not applied, and it was equivalent to the Young model. As w_2 was lowered, the application ratio of Rule 4 increased. In comparison with Figure 8, white regions were generated in the black spots, and instability increased and began pulsating. This phenomenon is attributed to Rule 4.

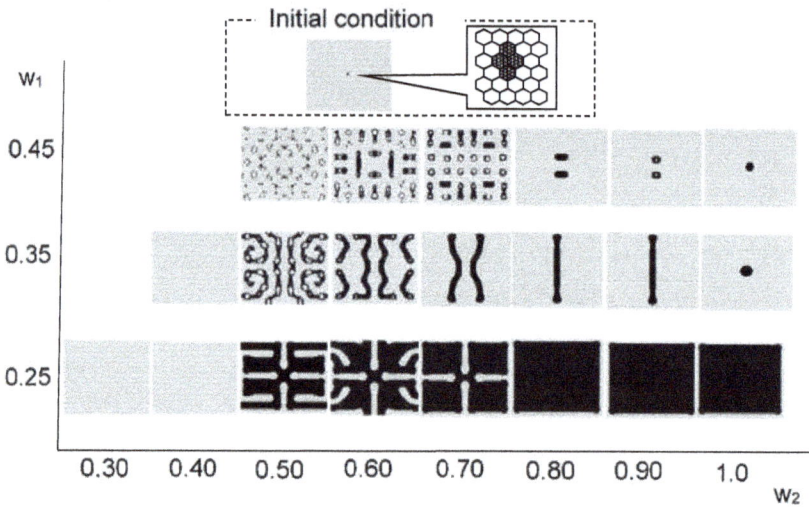

Figure 11. Results for the model with $s = 5$ and $t = 9$ in a hexagonal grid. The image of each w_1 and w_2 value is based on centrally arranged four black cells as shown in Figure 7. Instability became stronger as w_2 decreased. The initial central spotted pattern grew as an unstable wave in the field. When w_2 was relatively large, the central spot did not spread.

Figure 12 shows the time series results of the model with $s = 4$, $t = 8$, $w_1 = 0.40$, and $w_2 = 0.75$. In this figure, i shows the iteration number of the calculation. The video of Figure 12 can be viewed in the Video S1 of Supplementary Materials. As the initial condition, four state-1 cells (black) were set in the center of the field shown in Figure 7. The central spotted pattern was divided into two spots, and these divisions spread until the field was filled with spotted patterns. Figure 13 shows the results when the initial condition of state 1 was subjected to the same conditions as in Figure 12 ($s = 4$, $t = 8$, $w_1 = 0.40$, and $w_2 = 0.75$ in a hexagonal grid). When there was only one state 1 in the initial field, Rule 4 was applied around this state 1, which disappeared at the next step. If there are two or more states 1 as the initial state, it was observed that the state 1 survived, and pattern formation similar to Figure 12 occurred. In the case of several state 1s, due to the balance between Rules 3 and 4, complicated patterns were observed in two or three steps from the beginning of the calculation, and thereafter, a self-replicating pattern was observed.

In addition, Figure 14 shows the calculation results when increasing the black region as the initial condition, under the same conditions as in Figure 12 ($s = 4$, $t = 8$, $w_1 = 0.40$, and $w_2 = 0.75$ in a hexagonal grid). The point symmetrical initial shape tended to be a fixed pattern due to the balance of the rules. On the other hand, if the black area was larger than a certain size, Rule 1 was mostly applied, and disappeared at the second step. To produce a self-replicating pattern, an asymmetric shape must initially occur. Figure 15 shows the transition of the ratios of states 1 and 2, and the ratio of applied cells of Rule 4 per time step under the same conditions as in Figure 12 ($s = 5$, $t = 9$, $w_1 = 0.40$, and $w_2 = 0.75$). In 90 to 100 steps, the frequency of pattern division became high, and the application ratio of Rule 4 increased. Rule 4 generated a white area in the black spot patterns, and instability increased; hence, the symmetrical shape collapsed and became a trigger for division.

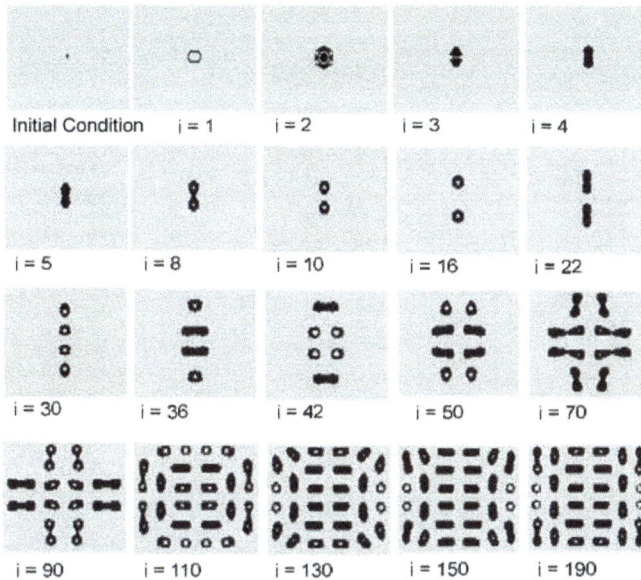

Figure 12. Results for the model with $s = 4$, $t = 8$, $w_1 = 0.40$, and $w_2 = 0.75$ in a hexagonal grid. As the initial condition, four state 1 cells (black) were set in the center of the field as shown in Figure 7. The central spotted pattern was divided into two spots, which then spread until the field was filled with spotted patterns.

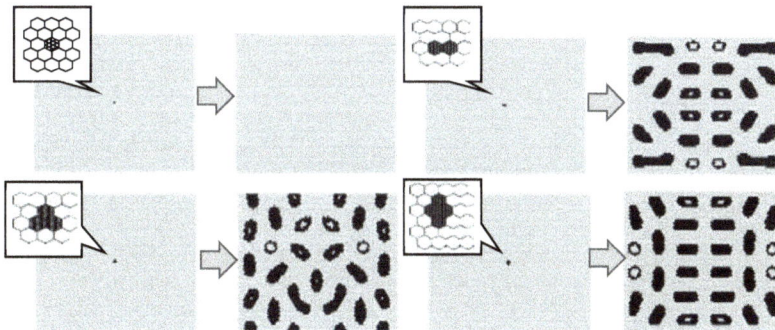

Figure 13. Results with the initial condition of state 1 under the same conditions as in Figure 12 ($s = 4$, $t = 8$, $w_1 = 0.40$, and $w_2 = 0.75$ in a hexagonal grid). If there is only one state 1 in the initial field, Rule 4 was applied at that location, which disappeared at the next step. If there are two or more state 1s as the initial state, state 1 was observed to survive, and a pattern formation similar to Figure 12 occurred.

Figures 16 and 17 show the results with an initial pattern of a symmetrical ring with state 1 s. In the case where the initial shape was symmetrical, the still pattern or the oscillation pattern often appeared in a wide range of w_1 and w_2. When $w_1 = 0.45$ and $w_2 = 0.55$, the ring pattern spread while dividing spots. In an opposite way, as w_2 was approached as w_1, the instability increased and the pattern disappeared.

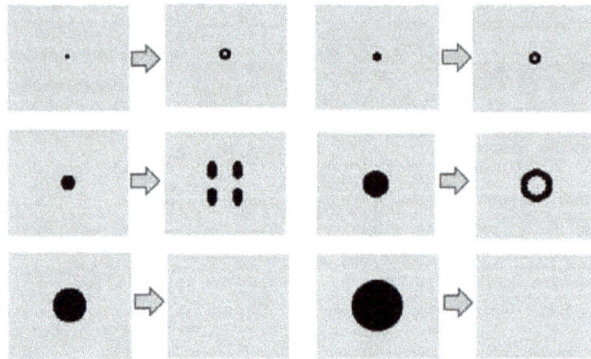

Figure 14. Calculation results, when increasing the black region as the initial condition, under the same conditions as in Figure 12 ($s = 4$, $t = 8$, $w_1 = 0.40$, and $w_2 = 0.75$ in a hexagonal grid). The point symmetrical initial shape tended to be a fixed pattern, due to the balance of each of the rules. On the other hand, if the black area is larger than a certain size, Rule 1 is mostly applied, and disappeared at the second step. To produce a self-replicating pattern, an asymmetric shape must initially occur.

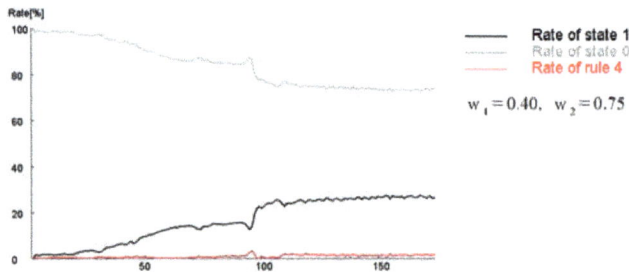

Figure 15. Transition of the ratios of states 1 and 2, and the ratio of the applied cells of Rule 4 per time step under the same conditions as in Figure 12 ($s = 5$, $t = 9$, $w_1 = 0.40$, and $w_2 = 0.75$). From 90 to 100 steps, the frequency of pattern division became high, and the application ratio of Rule 4 increased.

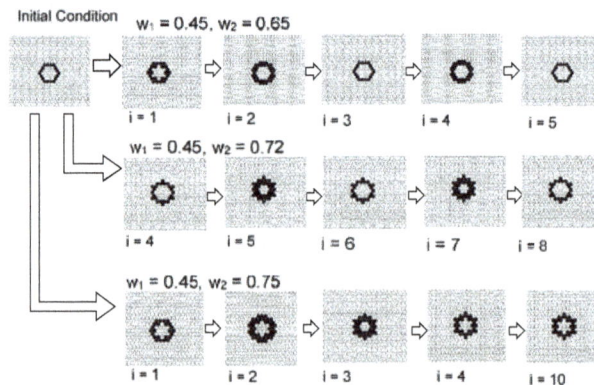

Figure 16. Results for the model with $s = 5$ and $t = 8$ in a hexagonal grid. The initial pattern was a symmetrical ring. When $w_1 = 0.45$ and $w_2 = 0.65$, the ring pattern became an oscillation pattern with period 2. When $w_1 = 0.45$ and $w_2 = 0.72$, the ring pattern likewise became an oscillation pattern with period 2. When $w_1 = 0.45$ and $w_2 = 0.75$, a still pattern was formed.

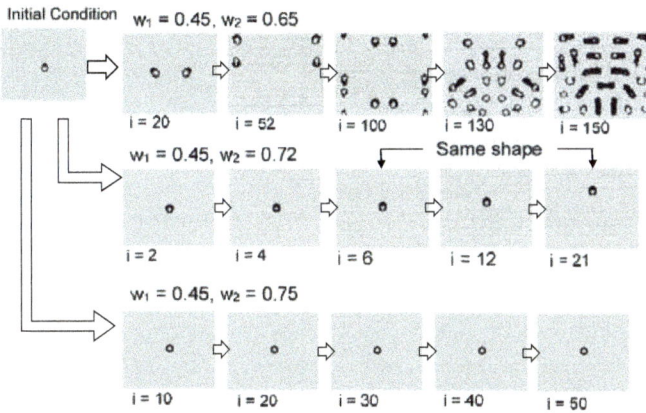

Figure 17. Results for the model with $s = 5$ and $t = 8$ in a hexagonal grid. The initial pattern was a symmetrical ring. When $w_1 = 0.45$ and $w_2 = 0.48$, the ring pattern disappeared. When $w_1 = 0.45$ and $w_2 = 0.55$, the ring pattern spread while breaking apart. When $w_1 = 0.45$ and $w_2 = 0.60$, still patterns were formed.

Figure 18 shows the initial pattern of a deformed ring with a thick upper boundary. When $w_1 = 0.45$ and $w_2 = 0.65$, the ring pattern was divided while moving. When $w_1 = 0.45$ and $w_2 = 0.72$, a constantly moving ring-like pattern formed. When $w_1 = 0.45$ and $w_2 = 0.75$, a single fixed symmetrical ring pattern formed. Thus, when the initial shape is asymmetric, a moving pattern or dividing pattern tends to be generated from the breaking of symmetry. As w_2 approaches 1, it approaches fixed patterns, and results in a pattern close to the Turing pattern.

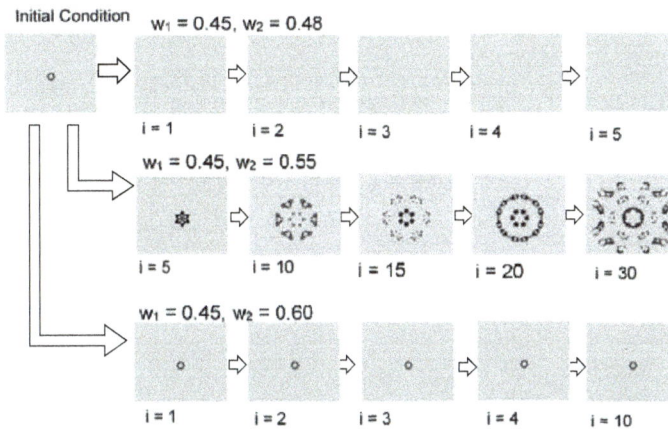

Figure 18. Results for the model with $s = 5$ and $t = 8$ in a hexagonal grid. The initial pattern was a deformed ring with a thick upper boundary. When $w_1 = 0.45$ and $w_2 = 0.65$, the ring pattern was divided while moving. When $w_1 = 0.45$ and $w_2 = 0.72$, a moving ring-like pattern formed. When $w_1 = 0.45$ and $w_2 = 0.75$, a single fixed symmetrical ring pattern formed.

4. Discussion

In this study, we constructed a model in which the focal state 1 survives only when a moderate rate of the number of state 1 cells surrounds the focal cell. This model created a central white part (state 0) within the black domain by changing parameter w_2. The patterns then became increasingly unstable over the calculation field, and this instability increased the rate of pattern emergence, including self-reproducing spot patterns and stripe patterns.

The self-reproducing spot pattern of Figure 12 seems to have some association with the Gray–Scott model, which is a type of RD model. Generally, the Gray–Scott model produces a self-reproducing spot pattern when the parameters are adjusted. In future work, we will investigate the relationship between the current model and the Gray–Scott model.

Particularly, in the case of the model with $s = 5$ and $t = 8$ shown in Figures 16–18, the result suggests the possibility of controlling self-organized patterns. When w_1 is increased, a spot shape can be generated, and by decreasing w_1, a stripe pattern can be grown from spotted seeds.

In the symmetric shape, when w_2 is large, it is stable in many cases. However, by making the value of w_2 small and making it close to w_1, the spotted pattern becomes unstable, and asymmetric shapes are born. Furthermore, these spots are moving or divided by the controlling w_2 parameter, as shown in Figure 19.

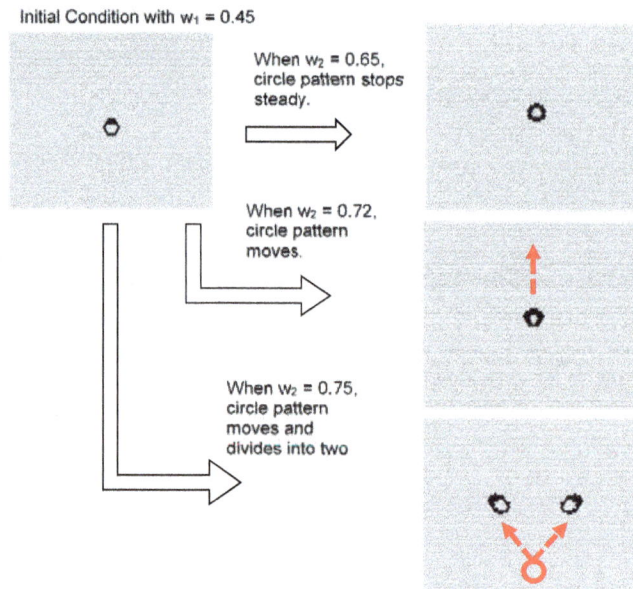

Figure 19. Possibility of self-organized ring patterns with one parameter. The results are a part of the model with $s = 5$ and $s = 8$ in a hexagonal grid, as shown in Figure 13. The ring pattern can be controlled with parameter w_2. When $w_2 = 0.65$, the ring pattern was divided while moving. When $w_2 = 0.72$, a moving ring pattern formed. When $w_2 = 0.75$, a fixed ring pattern formed.

The robustness of the initial conditions was found to have the following tendencies. (1) If state 1 is one, it disappears; (2) When state 1 is randomly arranged, if two or more states 1 are distributed in the range for counting the number of surrounding states, the development of the pattern is recognized. Next, depending on the degree of instability due to the w_2 parameter, it is decided whether the pattern develops spatially. Due to random initial placement, an identical pattern does not occur, but qualitatively similar patterns can be stably generated; (3) When several states 1 are gathered, complicated pattern changes occur in the first two or three steps due to the balance between Rules 3 and 4, and if a shape that is not point symmetric is subsequently generated, a pattern spreading in the space is generated. When a point symmetric shape is born, a circular or ring-fixed pattern is generated; (4) When the black (state 1) area is large, the black area disappears due to Rule 1.

In a future study, if CA rules can translate to a chemical reaction process, a control technology for nanomachines becomes feasible. Thus, the simple totalistic CA presented in this paper allows the emergence of a wide range of self-organization patterns to be observed.

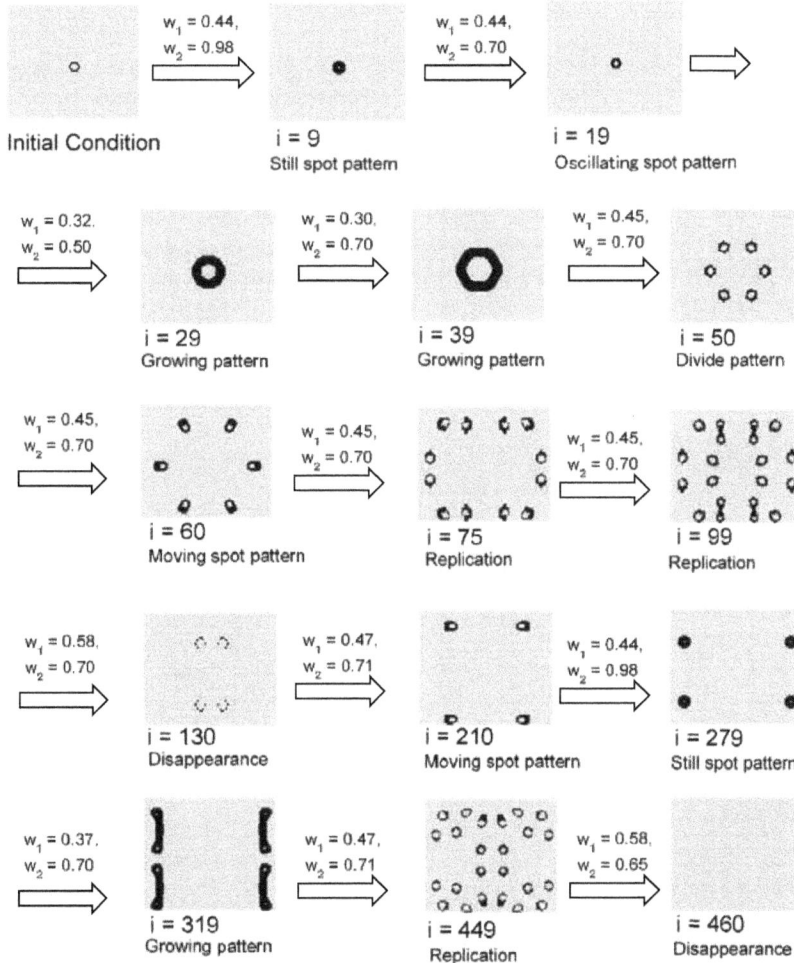

Figure 20. Calculation results in which w_1 and w_2 are changed over time at each step. The initial pattern was a hexagonal ring shape of state 1s, with $s = 5$ and $t = 8$ in a hexagonal grid. As in this result, it is possible to create changes by changing w_1 and w_2; when $i = 1$–9, $w_1 = 0.44$, and $w_2 = 0.98$, birth of annulus was observed; when $i = 11$–19, $w_1 = 0.44$, and $w_2 = 0.70$, an oscillating spot pattern was observed; when $i = 20$–29, $w_1 = 0.32$, and $w_2 = 0.50$, a growing ring pattern was observed; when $i = 30$–39, $w_1 = 0.30$, and $w_2 = 0.70$, a growing ring pattern was observed; when $i = 40$–99, $w_1 = 0.45$, and $w_2 = 0.70$, divide patterns and replication were observed; when $i = 100$–139, $w_1 = 0.58$, and $w_2 = 0.70$, pattern disappearance was observed; when $i = 140$–259, $w_1 = 0.47$, and $w_2 = 0.71$, moving spotted patterns were observed; when $i = 260$–279, $w_1 = 0.44$, and $w_2 = 0.98$, still circle patterns were observed; when $i = 280$–319, $w_1 = 0.37$, and $w_2 = 0.70$, a growing ring pattern was observed; when $i = 320$–449, $w_1 = 0.47$, and $w_2 = 0.71$, divide patterns and replication were observed; when $i = 460$–, $w_1 = 0.58$, and $w_2 = 0.65$, disappearance of all patterns was observed.

Based on these results, if the parameters w_1 and w_2 are changed at each time step, it is possible to generate a continuously changing pattern. Figure 20 shows a calculation case in which w_1 and w_2 are changed in time over time. The video of Figure 20 can be viewed in the Video S2 of Supplementary Materials. As in this result, it is possible to create changes by changing w_1 and w_2; birth of annulus -> growth -> division -> move -> decline -> growth -> annihilation.

Figure 21 shows the transition of the ratio of states 1 and 2, and the ratio of the applied cells of Rule 4 in each time step during the calculation of Figure 20. Rule 4 was applied more frequently when spot patterns replicated, as compared to when the spot moved or vibrated. Furthermore, when the pattern disappeared, the application frequency of Rule 4 increased. Thus, it is suggested that pattern stability can be controlled by changing the application ratio of Rule 4.

Since this research is still in its first stage, discussions are limited mainly to qualitative considerations, but in future work, it will be necessary to investigate further quantitative assessments. There is a possibility of finding an index like Langton's λ parameter with the evaluation of many calculation cases. Furthermore, since it has rules similar to the Game of Life, basically like the Game of Life, many trials are necessary to find the desired pattern. Slight changes in the initial value may develop into different patterns. In the future, it is necessary to make a catalog of pattern formation on the relation between the initial condition and patterns based on many calculation results.

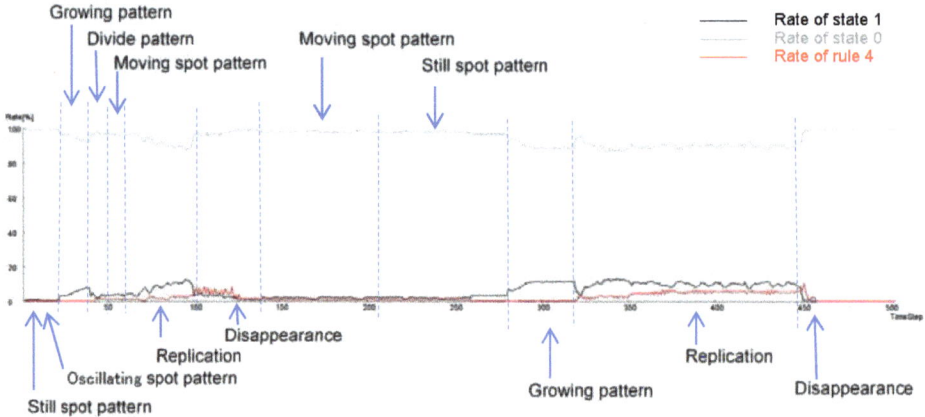

Figure 21. Transition of the ratio of states 1 and 2, and the ratio of the applied cells of Rule 4 in each time step during the calculation of Figure 20. Rule 4 was applied more frequently when spot patterns replicated than when the spot moved or vibrated. Furthermore, when the pattern disappeared, the application frequency of Rule 4 increased. Thus, it is suggested that pattern stability can be controlled by changing the application ratio of Rule 4.

5. Conclusions

To observe special patterns of emergent behavior that mimic biological systems, we developed a totalistic CA model with an activating inner area and inhibiting outer area. This is a CA model, yet it works in a way that is equivalent to that of RD approaches like the Turing pattern model. If a multiple CA model can be designed that allows the number of states to be increased, we may witness the emergence of more complex patterns, like those of living things.

Our model also has potential applications in the engineering field. For example, it may be possible to develop swarming robots that can work as a single unit or divide into groups. As shown in the result of Figure 13, spot-like regions can be formed in the space by this algorithm. If many robots are dispersedly arranged in this space, it becomes possible to gather by grouping the robots within this spot-like rounded area. Furthermore, using this algorithm, it is also possible to divide one spot area into two. Therefore, robots can be divided into two groups simply by paying attention so that the robot that was in the first spot area does not come out of the area.

The model can further be applied to information networks. If the transition rules can be applied to Internet of Things (IoT) devices through the Web, hierarchical structures of IoT devices may emerge. For example, if the IoT devices know their own position, this algorithm can be applied by transmitting tokens to the nearby devices. Then, the spotted structure can be formed in the network space of devices, and the devices can be divided into plural groups. Furthermore, if our method can be applied to microsystems, it may become possible to make a production process of self-organizing a microrobot from nanoparts automatically.

The model also produces Turing-like patterns using a simple algorithm and without the need to solve complex simultaneous differential equations. However, a weak point is the necessity to add up the state quantities within a certain area. While these calculations are simple to perform using electronic devices, it is difficult to see how an individual living cell in an organism would be able to calculate the states of distant cells. To apply this model to morphogenesis, therefore, further modification is needed.

Other outstanding questions include the mathematical relationship between this model and the RD equations, and the relationship with a Turing machine. It would also be interesting to consider the entropy of the model.

Supplementary Materials: The following are available online at http://www.mdpi.com/2072-666X/9/7/339/s1, Video S1: Video of Figure 12 (Results for the model with $s = 4$, $t = 8$, $w_1 = 0.40$, and $w_2 = 0.75$ in a hexagonal grid); Video S2: Video of Figure 20 (Calculation results in which w_1 and w_2 are changed over time at each step).

Funding: This research was supported by grants from the Project of the NARO (National Agriculture and Food Research Organization) Bio-oriented Technology Research Advancement Institution (the special scheme project on regional developing strategy).

Conflicts of Interest: The authors declare no conflict of interest.

References

1. Turing, A.M. The Chemical Basis of Morphogenesis. *Philos. Trans. R. Soc. Lond. B Biol. Sci.* **1952**, *237*, 37–72. [CrossRef]
2. Gierer, A.; Meinhardt, H.A. A Theory of Biological Pattern Formation. *Kybern. Biol. Cybern.* **1972**, *12*, 30–39. [CrossRef]
3. Gray, P.; Scott, S. Autocatalytic Reactions in the Isothermal Continuous Stirred Tank Reactor. *Chem. Eng. Sci.* **1984**, *38*, 1087–1097. [CrossRef]
4. Neumann, J. *Theory of Self-Replicating Automata*; University of Illinois Press: Champaign, IL, USA, 1966.
5. Mange, D.; Stauffer, A.; Peparodo, L.; Tempesti, G.G. A Macroscopic View of Self-Replication. *Proc. IEEE* **2004**, *92*, 1929–1945. [CrossRef]
6. Codd, E.F. *Cellular Automata*; Academic Press: Cambridge, MA, USA, 1968.
7. Gardner, M. Mathematical Games–The fantastic combinations of John Conway's new solitaire game "life". *Sci. Am.* **1970**, *223*, 120–123. [CrossRef]
8. Wolfram, S. Universality and Complexity in Cellular Automata. *Physics D* **1984**, *10*, 1–35. [CrossRef]
9. Langton, C. (Ed.) *Artificial Life*; Addison-Wesley: Boston, MA, USA, 1989; pp. 1–48.
10. Byl, J. Self-Reproduction in Small Cellular Automata. *Physics D* **1989**, *34*, 295–299. [CrossRef]
11. Reggia, J.A.; Lohn, J.D.; Chou, H.H. Self-Replicating Structures: Evolution, Emergence, and Computation. *Artif. Life* **1998**, *4*, 283–302. [CrossRef] [PubMed]
12. Ishida, T. Simulating Self-reproduction of Cells in a Two-dimensional Cellular Automaton. *J. Robot Mechatron.* **2010**, *22*, 669–676. [CrossRef]
13. Young, D.A.A. Local Activator-Inhibitor Model of Vertebrate Skin Patterns. *Math. Biosci.* **1984**, *72*, 51–58. [CrossRef]
14. Madore, B.F.; Freedman, W.L. Computer Simulation of the Belousov-Zhabotinsky reaction. *Science* **1983**, *222*, 615–616. [CrossRef] [PubMed]

15. Gerhardt, M.; Schuster, H. A Cellular Automaton Describing the Formation of Spatially Ordered Structures in Chemical Systems. *Phys. D* **1989**, *36*, 209–221. [CrossRef]
16. De Boer, R.J.; Hogeweg, P.; Perelson, A.S. Growth and Recruitment in the Immune Network. In *Theoretical and Experimental Insights into Immunology*; Springer-Verlag: New York, NY, USA, 1992; Volume 66, pp. 223–247.
17. Celadaa, F.; Seidenb, P.E. A Computer Model of Cellular Interactions in the Immune System. *Immunol. Today* **1992**, *13*, 56–62. [CrossRef]
18. Moreira, J.; Deutsch, A. Cellular Automaton Models of Tumor Development: A Critical Review. *Adv. Complex Syst.* **2002**, *5*, 247–269. [CrossRef]
19. Moore, J.H.; Hahn, L.W. A Cellular Automata-based Pattern Recognition Approach to Identifying Gene-gene and Gene-environment Interactions. *Am. J. Hum. Genet.* **2000**, *67*, 52.
20. Pytte, E.; Grinstein, G.; Traub, R.D. Cellular Automaton Models of the CA3 Region of the Hippocampus. *Network* **1991**, *2*, 149–167. [CrossRef]
21. Kaplan, D.T.; Smith, J.M.; Saxberg, B.E.H.; Cohen, R.J. Nonlinear Dynamics in Cardiac Conduction. *Math. Biosci.* **1988**, *90*, 19–48. [CrossRef]
22. Markus, M.; Hess, B. Isotropic Cellular Automaton for Modeling Excitable Media. *Nature* **1990**, *347*, 56–58. [CrossRef]
23. Langton, C.G. Computation at the edge of chaos: Phase transitions and emergent computation. *Physics D* **1990**, *42*, 12–37. [CrossRef]
24. Tokihiro, T.; Takahashi, D.; Matsukidaira, J.; Satsuma, J. From Soliton Equations to Integrable Cellular Automata through a Limiting Procedure. *Phys. Rev. Lett.* **1996**, *76*, 3247–3250. [CrossRef] [PubMed]
25. Sipper, M. Fifty years on Self-Replicating; An Overview. *Artif. Life* **1998**, *4*, 237–257. [CrossRef] [PubMed]
26. Adamatzky, A.; Martínez, G.J.; Carlos, J.; Mora, S.T. Phenomenology of reaction-diffusion binary-state cellular automata. *Int. J. Bifurc. Chaos* **2006**, *16*, 2985–3005. [CrossRef]
27. Dormann, S.; Deutsch, A.; Lawniczak, A.T. Fourier analysis of Turing-like pattern formation in cellular automaton models. *Future Gener. Comput. Syst.* **2001**, *17*, 901–909. [CrossRef]
28. Tsai, L.L.; Hutchison, G.R.; Peacock-Lopez, E. Turing patterns in a self-replicating mechanism with a self-complementary template. *J. Chem. Phys.* **2000**, *113*, 2003–2006. [CrossRef]
29. Nakamasu, A.; Takahashi, M.; Kondo, S. Interactions between Zebrafish Pigment Cells Responsible for the Generation of Turing Patterns. *Proc. Natl. Acad. Sci. USA* **2009**, *106*, 8429–8434. [CrossRef] [PubMed]

micromachines

MDPI

Article

Artificial Cochlear Sensory Epithelium with Functions of Outer Hair Cells Mimicked Using Feedback Electrical Stimuli

Tetsuro Tsuji [ID], Asuka Nakayama, Hiroki Yamazaki and Satoyuki Kawano *

Graduate School of Engineering Science, Osaka University, Toyonaka, Osaka 560-8531, Japan;
tsuji@me.es.osaka-u.ac.jp (T.T.); nakayama@bnf.me.es.osaka-u.ac.jp (A.N.);
yamazaki@bnf.me.es.osaka-u.ac.jp (H.Y.)
* Correspondence: kawano@me.es.osaka-u.ac.jp; Tel.: +81-6-6850-6175

Received: 24 April 2018; Accepted: 24 May 2018; Published: 30 May 2018

Abstract: We report a novel vibration control technique of an artificial auditory cochlear epithelium that mimics the function of outer hair cells in the organ of Corti. The proposed piezoelectric and trapezoidal membrane not only has the acoustic/electric conversion and frequency selectivity of the previous device developed mainly by one of the authors and colleagues, but also has a function to control local vibration according to sound stimuli. Vibration control is achieved by applying local electrical stimuli to patterned electrodes on an epithelium made using micro-electro-mechanical system technology. By choosing appropriate phase differences between sound and electrical stimuli, it is shown that it is possible to both amplify and dampen membrane vibration, realizing better control of the response of the artificial cochlea. To be more specific, amplification and damping are achieved when the phase difference between the membrane vibration by sound stimuli and electrical stimuli is zero and π, respectively. We also demonstrate that the developed control system responds automatically to a change in sound frequency. The proposed technique can be applied to mimic the nonlinear response of the outer hair cells in a cochlea, and to realize a high-quality human auditory system.

Keywords: artificial cochlea; MEMS; piezoelectric material; outer hair cell

1. Introduction

Hearing is important to infants in terms of acquiring language and sentiment education. However, one in a thousand new-born infants suffers congenital deafness. Meanwhile, the ability of adults to hear diminishes gradually and never recovers spontaneously. There is thus an increasing demand for artificial cochleae, which are devices that support patients with sensorineural hearing loss. The present study deals with the development of a novel technique that allows the design of a higher-quality artificial cochlea to be combined with previous devices developed by one of the authors [1–3].

The vibration of sound is collected and amplified by outer and middle ears, and transmitted to the inner ear. The frequency of sound is then distinguished by cochlear epithelium in the inner ear [4], and the vibration of sound is converted into electrical signals by inner hair cells to stimulate nerves, which are sensory cells on the epithelium. Because damaged hair cells do not regenerate spontaneously [5], cochlear implants [6,7] are used to effectively overcome sensorineural hearing loss due to cochlear damage . However, commercially available cochlear-implant devices require patients to wear external equipment such as microphones, sound processors, and batteries. To remove such burdens on patients and to increase patients' quality of life, a self-contained artificial cochlea has been proposed by one of the authors and their colleagues [1–3]. Recent advances in the fabrication technologies of micro-electro-mechanical systems have allowed the development of an artificial cochlear epithelium without an external power supply [1,2,8–19]. Electrical signals in these devices are generated by the deformation of a trapezoidal piezoelectric membrane induced by sound stimuli. Such a mechanical generation of electrical signals has been proposed as an energy harvester [20] or flow sensor [21] in biological systems, and its performance as a micro-electro-mechanical-system-fabricated acoustic transducer has been quantitatively investigated [22,23]. Moreover, a method of fixing the device in the cochlea has also been proposed by one of the authors [24] to show the feasibility of the artificial cochlear epithelium made of piezoelectric materials.

In addition to having frequency selectivity and converting vibration to electrical signals, the cochlea amplifies signals through active feedback [25–27]. In fact, the cochlear basilar membrane has a nonlinear response to sound stimuli, and this response is attributed to the functions of the outer hair cells in the organ of Corti [28–33]; that is, the outer hair cells have the function of controlling the sound level to be recognized. To be more specific, a weak sound input is amplified and a strong input is damped. Thanks to these nonlinear responses, human hearing has a wide dynamic range from 0 to 120 dB in amplitude, although the range of displacement of the cochlear epithelium is from 0.1 to 10 nm [30]. Since the nonlinear response is partly realized by the mechanical feature of the basilar membrane [30], it is important to mimic this feature in engineering the artificial cochlea. In fact, an attempt to mimic the nonlinear response has been made using a cantilever model with dimensions of 470 × 38 mm [34]. It is shown in Reference [34] that a cubic damping term with respect to the cantilever velocity in the control signal realizes the nonlinear response of the cantilever velocity against a disturbance. Moreover, the nonlinear response had similar characteristics with human hearing.

The present paper reports the development of an artificial cochlear epithelium which controls the local vibration according to sound stimuli, on the basis of our preceding study [2]. To be more specific, the preceding study [2] mimicked the function of the basilar membrane and inner hair cells, while the present study also mimics the function of outer hair cells. The development of such a control technique is expected to produce a device that realizes the nonlinear response. Moreover, the control technique can be used to better understand the cochlear mechanism by providing an experimental method that replicates the cochlear behavior.

The artificial cochlear epithelium in the present study is a trapezoidal piezoelectric membrane with patterned Al electrodes [2], where each electrode has two small Al plates for the electrical output and feedback input. The resonance frequency of the device varies locally according to the longitudinal position, achieving the function of frequency selectivity. The patterned electrodes have two functions. One is to recognize sound stimuli through the piezoelectric effect induced by the local deformation of the membrane, while the other is to control local vibration through the inverse piezoelectric effect by applying external electrical stimuli. It is found that the local membrane vibration can be amplified or damped by choosing appropriate phase differences between the electrical and sound stimuli. Using these amplification and damping controls, the nonlinear response that partly realizes the wide dynamic range of human hearing is obtained. Such techniques for recognition and control can be applied to the development of the artificial cochlear epithelium, which mimics functions of the basilar membrane, inner hair cells, and outer hair cells. These functions are needed to achieve the high performance of the human cochlea, having a wide range frequency selectivity from 20 to 20,000 Hz and a wide dynamic range from 0 to 120 dB.

2. Experimental Method

2.1. Fabrication of the Artificial Cochlear Epithelium

We first give an overview of the developed device. We use a piezoelectric membrane having a fixed boundary conditions with a trapezoidal shape as shown in Figure 1a. To control the vibration of the membrane, we apply electrical stimuli to patterned electrodes fabricated on the piezoelectric membrane. The patterned electrodes can also be used to recognize the vibration amplitude. Note that in previous studies [1,2], the electrodes were used to generate the electrical signals to stimulate nerves, mimicking the function of the inner hair cells. In the present study, however, the electrodes are used for feedback control to also mimic the function of the outer hair cells. The vibrations induced by sound and electrical stimuli are expected to be superposed. The piezoelectric material used in this study is a polyvinylidene difluoride (PVDF), because PVDF is biologically compatible [8]. We describe the fabrication method and details of the device in the following.

The device has patterned Al electrodes, as shown in Figure 1a,b. Electrodes have two functions, and are referred to as recognition and control electrodes accordingly. Six pairs of recognition and control electrodes are fabricated on the PVDF membrane. The backside of the PVDF membrane is completely covered by Al deposition and is electrically grounded. The number of electrode pairs was chosen so that local amplitude control can be demonstrated while minimizing the complexity of the device.

The Cartesian coordinate system is defined as in Figure 1a, and the x-coordinates of the center of the i-th control electrode, $x^{(i)}$ ($i = 1, 2, \cdots, 6$), are set as $x^{(1)} = 5$ mm, $x^{(2)} = 9$ mm, $x^{(3)} = 13$ mm, $x^{(4)} = 17$ mm, $x^{(5)} = 21$ mm, and $x^{(6)} = 25$ mm. The dimensions of the recognition electrodes (0.5×1.0 mm centered at $y = 0$ mm) are the same as those in our previous study [2], while those of control electrodes are smaller (0.2 mm) to achieve a better localization of electrical stimuli. The distance between recognition and control electrodes is 0.2 mm. This value of distance was chosen to be small in order to apply electrical stimuli near the recognition electrodes, but the distance should be large enough to achieve a stable electrode-patterning process on the PVDF membrane. The electrode pattern is designed so that geometrical symmetry is preserved but the main feature of the recognition electrode is the same as that in our previous study [2]. To be more precise, the pattern is symmetric with respect to the x-axis, except for the small gap indicated in the magnified figure of Figure 1a. The electrodes with diagonal lines are prepared to have geometrical symmetry, and were not used in this study.

The device was fabricated by attaching the PVDF membrane to a stainless-steel plate with a trapezoidal slit using double-sided tape (No. 500, Nitto Denko, Osaka, Japan). The fabrication process is summarized in the Appendix. The method of assembly was carefully chosen so that the device will not degrade with time. The slit has length $L_x = 30$ mm in the x-direction, and short and long sides in the y-direction are $L_{y,\text{short}} = 2$ mm and $L_{y,\text{long}} = 4$ mm in length, as shown in Figure 1a. The length in the y-direction of the trapezoidal membrane is therefore expressed as $l(x) = L_{y,\text{short}} + (L_{y,\text{long}} - L_{y,\text{short}})(x/L_x)$. These values are the same as those in Reference [2].

2.2. Experimental Setup

An overview of the experimental setup is shown in Figure 1c. The data acquisition (DAQ) system consists of a controller (NI PXIe-1082, NI PIXIe-8840, National Instruments, Austin, TX, USA), multichannel analog output module (NI PIXIe-6738, National Instruments, Austin, TX, USA) for sound and electrical stimuli, and a vibration module (NI PIXIe-6738, National Instruments, Austin, TX, USA) for data analysis. The device was fixed on an x–y auto-stage, which was also controlled by the DAQ system. A laser Doppler vibrometer (LDV; AT3600, AT0023, Graphtec, Tokyo, Japan) was used to measure the vibration of the device. The laser was irradiated through a mirror system (Figure 1d), where a camera (DCC1645C, Thorlabs, Newton, NJ, USA) and light-emitting diode as a light source (LEDD18/M565L3, Thorlabs, Newton, NJ, USA) were installed to precisely observe the position of the laser (i.e., the measuring point). The electrical signal V_s from the DAQ system was magnified by an amplifier (HSA4-14, NF, Yokohama, Japan) and applied to a speaker (FX102, Fostex, Tokyo, Japan). Here, V_s was set to $V_s = \bar{V}_s \sin(2\pi f_s t)$, where $\bar{V}_s$, f_s, and t are respectively the voltage amplitude, frequency of sound stimuli, and time variable. A microphone (377C01, 426B03, 480C02, Piezotronics, Depew, NY, USA) was used to evaluate the magnitude of the sound stimuli, P_s (dB SPL). To measure P_s near the membrane, we fabricated the acyclic part as shown in Figure 1e. Sound came from A in Figure 1e through a tube connected to a speaker, and passed to B (the device) and C (the microphone). The distance between point B and the membrane was set to less than 4 mm. The electrical signal $V_e^{(i)}$ from the DAQ system was applied to the *i*-th control electrode, where

$$V_e^{(i)} = \bar{V}_e^{(i)} \sin(2\pi f_e^{(i)} t + \phi^{(i)}), \quad (i = 1, 2, \cdots, 6). \tag{1}$$

The parameter $\phi^{(i)}$ is the phase difference between the signals for sound and electrical stimuli. Details are given in Section 3.2. The electrical outputs $V_{rec}^{(i)}$ from the *i*-th recognition electrodes and their amplitudes $\bar{V}_{rec}^{(i)}$ were stored in the DAQ system.

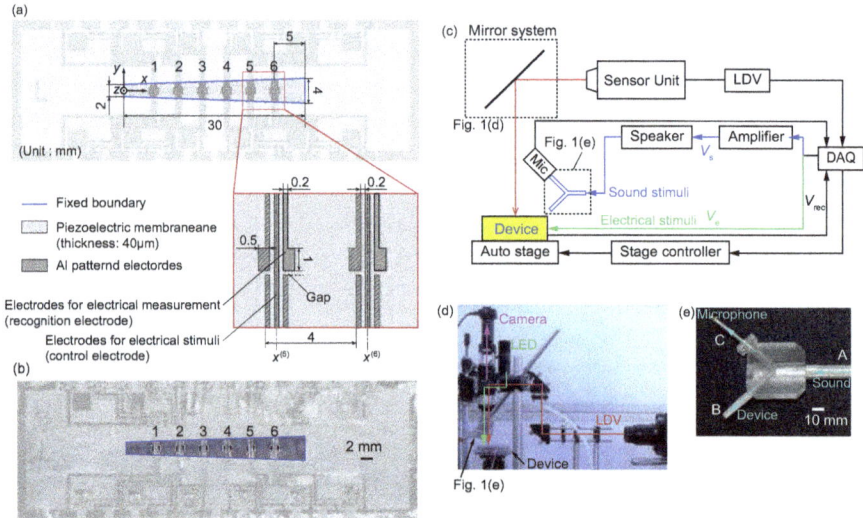

Figure 1. (**a**) Schematic and (**b**) photograph of the artificial cochlear epithelium fabricated in this paper. The patterned electrodes were fabricated on a polyvinylidene difluoride (PVDF) piezoelectric membrane with thickness of 40 μm; (**c**) Overview of the experimental setup. The multichannel analog output module in the data acquisition (DAQ) system was used to apply the sound and electrical stimuli with a desired phase difference. The DAQ system was also used to analyze the electrical output from the device and the results of laser Doppler vibrometer (LDV) measurement. These analyzed data were used in feedback control; (**d**) Details of the mirror system in panel (c) to observe the exact position of the laser for LDV measurement; (**e**) Photograph of an acyclic part used to fix the microphone.

3. Results and Discussion

3.1. Frequency Selectivity

First of all, we will briefly present the function of frequency selectivity of the present device. More detailed investigations of the function of frequency selectivity (e.g., a comparison with the Wentzel–Kramers–Brillouin (WKB) asymptotic solution and the effect of the surrounding fluid) were conducted in our previous study [2]. Figure 2 shows the amplitude distribution of membrane vibration in the *x*–*y* plane. Here, the vibration was induced by sound stimuli with f_s = 4.6, 5.6, 6.6, and 8.2 kHz. These frequencies are close to the resonant frequencies of second, third, fourth, and fifth electrodes, respectively, which will be presented in Section 3.2.1. Contours show the relative amplitude normalized by the maximum value for each f_s. The resonant positions, at which the maximum value is obtained, are indicated by white dots. For each f_s, the resonant positions were different, and we thus conclude that the present device has frequency selectivity. Owing to the vibration mode in the *x*-direction and/or higher frequency modes, a couple of peaks were observed for each f_s. This tendency became prominent as f_s increased and was the same as that of the previous device [2], but is not observed in the mammalian cochlear epithelium [33]. The duplication of the full details of the vibration modes observed in the mammalian cochlear epithelium will require the geometry and boundary condition to be mimicked. This difficulty is inherent to the membrane vibration, and its solution is our ultimate goal of future research. The development of a device that has simpler vibration modes and a clear one-to-one relation between the resonant position and frequency will help provide a solution.

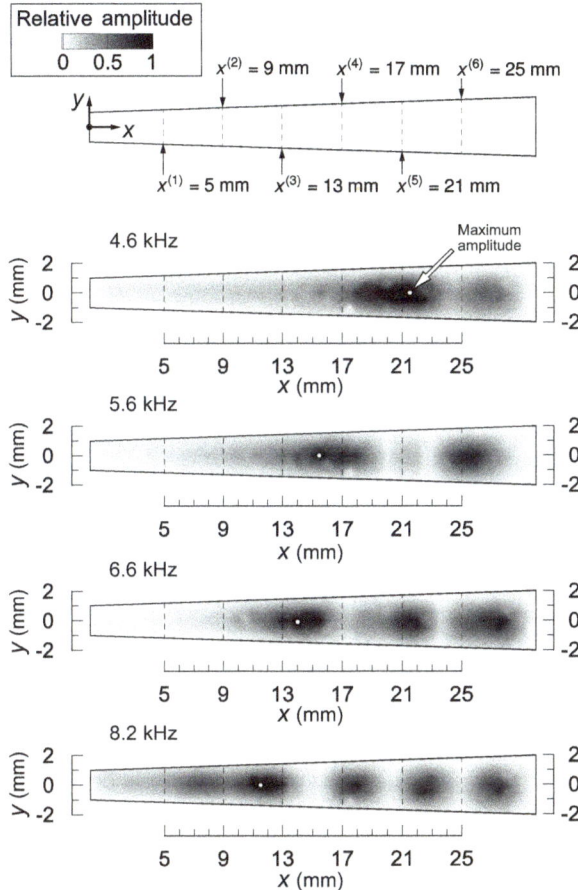

Figure 2. Amplitude distribution of membrane vibration in the x–y plane induced by sound stimuli with f_s = 4.6, 5.6, 6.6, and 8.2 kHz. Contours show the relative amplitude normalized by the maximum value for each f_s. The resonant positions, at which the maximum value was obtained, are indicated by white dots. For each f_s, the resonant positions were different, and we thus conclude that the present device had frequency selectivity. Owing to the vibration mode in the x-direction, a couple of peaks were observed for each f_s. This tendency is the same as that observed for a previous device [2] developed by one of the authors.

3.2. Local Vibration Control Using Electrical Stimuli

This section describes the results of local vibration control using electrical stimuli. Section 3.2.1 determines parameters needed for vibration control, while Sections 3.2.2 and 3.2.3 present examples of local vibration control.

3.2.1. Search for Resonant Frequencies and Control Parameters

We first show that electrical stimuli induce vibration of the membrane and can be used to determine resonant frequencies $f_{res}^{(i)}$ at position $x^{(i)}$ of the control electrodes by changing $f_e^{(i)}$ in Equation (1) continuously. Note that we measured the vibration of the membrane using an LDV and obtained the amplitude a_f for the frequency component f through Fourier analysis.

Figure 3a shows the amplitude of membrane vibration on the fifth electrode ($x = 21$ mm) induced by electrical stimuli applied to that electrode with different frequencies 4 kHz $\leq f_e^{(5)} \leq$ 6 kHz and $\bar{V}_e^{(5)} = 5$ V. The amplitude reached a maximum at $f_e^{(i)} = 4.66$ kHz, and we thus conclude that the resonant frequency at the position of the fifth electrode was $f_{res}^{(5)} = 4.66$ kHz. Before going into further detail, we will compare the experimental results with a theoretical prediction. Assume that the trapezoidal membrane can be considered as sequential beams of length $l(x)$; that is, we neglect the vibration mode in the x-direction. Then, according to the linear theory of transverse vibration of a beam with fixed ends [35], we obtain the first-order resonant frequency as $f_{res}^{(\text{theory})} = (22.37/2\pi)(EJ/\rho A)^{1/2}/l^2(x)$, where E is the modulus of elasticity, ρ the mass density, $A = hl(x)$ the cross-sectional area with h the thickness of the membrane, and $J = h^3 l(x)/12$ the moment of inertia of the cross section about the x-axis. In other words, $C(x) = f_{res}^{(\text{theory})} l^2(x)$, which is introduced for notational convenience, should be independent from x; that is, $C(x) = C_0 = (22.37/2\pi)(Eh^2/12\rho)^{1/2}$ m$^2\cdot$s^{-1}. The experimental values of $C^{(i)} = C(x^{(i)})$ for positions $x^{(i)}$ of the electrodes used in the present study were obtained as $C^{(i)} = (5.36 \pm 0.16) \times 10^{-2}$ m$^2\cdot$s^{-1}. The standard deviation was 3%, and we thus conclude that the linear theory predicts the resonant frequency of the electrodes well. Although the above discussion using beam theory is simple, the main feature of the device's frequency selectivity is retained. Further details such as vibration mode and phase will be investigated in future work.

In Figure 3a, the amplitude was 3 nm at most, and it will be seen throughout the paper that the amplitude was on the order of 1 nm. Note that the present LDV equipment ensured a measurement of amplitude on the order of 1 nm for the frequency range from 1 to 100 kHz, which includes the frequencies treated in the present paper. According to the linear theory, we assumed that the sound-stimuli-induced vibration and the electrical-stimuli-induced vibration can be superposed when both stimuli are applied. The resonant frequencies of the device varied slightly when the device was unmounted from the experimental system and mounted again later. Therefore, the measurements of $f_{res}^{(i)}$ were checked and adjusted again before each experiment. For the second, third, fourth, and fifth electrodes (which are mainly discussed in the present paper), the average resonant frequencies were, respectively, $f_{res}^{(2)} = 8.13 \pm 0.04$ kHz, $f_{res}^{(3)} = 6.45 \pm 0.19$ kHz, $f_{res}^{(4)} = 5.69 \pm 0.14$ kHz, and $f_{res}^{(5)} = 4.55 \pm 0.15$ kHz. Note that the most important frequency range in daily life is between 1 kHz and 3 kHz [36], and is lower than the resonant frequencies $f_{res}^{(i)}$ obtained for the present device. Because the resonant frequency is proportional to the membrane thickness h as explained above, a device with a thinner membrane will be suitable for a lower hearing range, and this will be a topic of our future work. The C values for these four electrodes are given as $C^{(2)} = 5.49 \times 10^{-2}$ m$^2\cdot$s^{-1}, $C^{(3)} = 5.30 \times 10^{-2}$ m$^2\cdot$s^{-1}, $C^{(4)} = 5.59 \times 10^{-2}$ m$^2\cdot$s^{-1}, and $C^{(5)} = 5.26 \times 10^{-2}$ m$^2\cdot$s^{-1}. Note that these four electrodes are suitable for the validation of the proposed technique in the present paper because the first and sixth electrodes are close to the fixed ends in the x-direction and are affected by boundary conditions.

For vibration control using electrical stimuli, we must choose appropriate parameters $\bar{V}_e^{(i)}$, $f_e^{(i)}$, and $\phi^{(i)}$ contained in Equation (1) according to the sound stimuli. Here, as a test case, we determined these parameters such that the vibration of the membrane at the fifth electrode was amplified twice when sound stimuli at the resonant frequency, $f_s = f_e^{(5)}$, were applied. To be more precise, V_s was chosen so that $a_f = 2.5$ nm when $\bar{V}_e^{(5)} = 0$ V, and $\bar{V}_e^{(5)}$ was chosen so that $a_f = 5.0$ nm at most when both sound and electrical stimuli were applied. The frequency of electrical stimuli $f_e^{(5)}$ should be the same as f_s, otherwise no clear amplification or damping is expected in the present linear regime. We finally chose $\phi^{(5)}$ from the measurement. Figure 3b presents the amplitude a_f for different phases $\phi^{(5)}$ when both sound and electrical stimuli were applied. It is clear that the amplitude was at a maximum and $a_f \approx 5$ nm for a certain phase $\phi^{(5)} = \phi_{max}^{(5)} \approx 1.8\pi$, while the amplitude was at a minimum and $a_f \approx 0$ nm for another phase $\phi^{(5)} = \phi_{min}^{(5)} \approx 0.8\pi$. In this manner, we can determine the phase difference of electrical stimuli for each control electrode to amplify/dampen the vibration. Note that the vibration of the membrane induced only by the sound stimuli and that only by the electrical stimuli had the same phase when there was amplification and had the opposite phase when there was damping.

Finally, we make some comments on the above experiments. It would be better if we could have predicted the values of ϕ_{max} and ϕ_{min} by measuring all the phase shifts in the experimental setup (e.g., by measuring the distance from the speaker and the phase shift in the electrical circuits). However, because the speaker has a typical dimension of 100 mm and it is difficult to define the distance from the speaker to the device, we measured and determined ϕ_{max} and ϕ_{min} experimentally. We could have attached an oscillator to the stainless-steel plate instead of the sound stimuli to induce the vibration of the membrane and to measure the piezoelectric output. However, this would have complicated the LDV analysis because we would have to distinguish the vibration of the membrane and that of the stainless-steel plate. Moreover, experiments with sound stimuli are more appropriate because our device is applied to a hearing device.

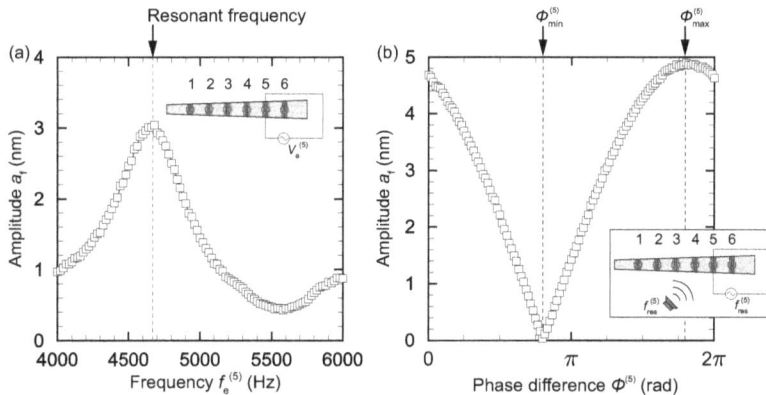

Figure 3. (a) Vibration amplitude at the fifth electrode ($x = 21$ mm) for electrical stimuli applied to the fifth electrode with frequency $f_e^{(5)}$. The resonant frequency can be determined by sweeping the frequency $f_e^{(5)}$ in Equation (1); (b) Phase dependence on the vibration amplitude at the fifth electrode under sound stimuli with $f_s = f_{res}^{(5)}$ and electrical stimuli with $f_e^{(5)} = f_{res}^{(5)}$. The amplitude is at a maximum and $a_f \approx 5$ nm for a certain phase $\phi^{(5)} = \phi_{max}^{(5)} \approx 1.8\pi$, while the amplitude is at a minimum and $a_f \approx 0$ nm for an another phase $\phi^{(5)} = \phi_{min}^{(5)} \approx 0.8\pi$. The appropriate choice of $\phi^{(i)}$ leads to amplification or damping control of the membrane vibration.

3.2.2. Improvement of The Response of The Device through Vibration Control

This section presents the ability to control the vibration of the membrane by applying electrical stimuli. The amplitude was measured by the LDV for the entire range of x (i.e., 0 mm $\leq x \leq$ 30 mm). In the following, $\tilde{V}_e^{(i)}$ is determined such that the electrical stimuli amplify the vibration by a factor of two with $\phi^{(i)} = \phi_{max}^{(i)}$. To be more precise, $\tilde{V}_e^{(i)}$ was chosen so that $a_f|_{x=x^{(i)}} = 2a_f'|_{x=x^{(i)}}$ with $\phi^{(i)} = \phi_{max}^{(i)}$, where $a_f'|_{x=x^{(i)}}$ is the amplitude at $x = x^{(i)}$ with all $\tilde{V}_e^{(i)}$ set to zero. The goal of this section is to amplify the vibration at the i_t-th electrode. We therefore refer to the i_t-th electrode as a target electrode. Resonant frequencies $f_{res}^{(i)}$ and phases $\phi_{min}^{(i)}, \phi_{max}^{(i)}$ were measured for all electrodes, prior to the experiments described below.

(A) Only the sound stimuli with $f_{res}^{(i_t)}$ is applied.

(B) In addition to protocol (A), electrical stimuli are applied to the i_t-th electrode with $\phi^{(i_t)} = \phi_{max}^{(i_t)}$.

(C) In addition to protocol (B), electrical stimuli are applied to the $(i_t \pm 1)$-th electrode with $\phi^{(i_t \pm 1)} = \phi_{min}^{(i_t \pm 1)}$.

Protocol (A) leads to the usual response of the device to sound stimuli. Protocol (B) tries to amplify the vibration at the resonant position. However, amplifying the vibration at the resonant position may lead to amplification of the vibration at off-resonant positions. Protocol (C) is similar to protocol (B), but the additional electrical stimuli suppress the vibration of off-resonant positions close to the resonant position, leading to better frequency selectivity. We set $i_t = 3, 4$, and 5 and carried out each experiment five times.

The results for $i_t = 3, 4$, and 5 are respectively presented in Figures 4–6. In each figure, panel (a) shows the amplitude distribution for the entire range of x while panel (b) is the magnified view around $x^{(i_t)}$. The plots show averages for five trials with the error bars representing standard deviations.

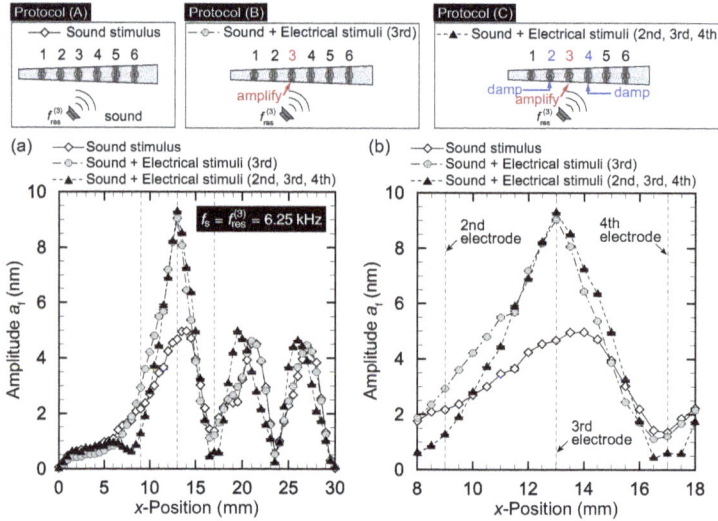

Figure 4. Vibration amplitude distribution for protocol (A): only sinusoidal sound stimuli with $f_s = 6.25$ kHz; protocol (B): sound and electrical stimuli applied to the third electrode; and protocol (C): sound and electrical stimuli applied to the second to fourth electrodes. (a) Entire view and (b) magnified version near $x^{(3)}$. Protocol (B) succeeded in amplifying a_f at resonant position $x^{(3)} = 13$ mm. In addition to the amplification in Protocol (B), Protocol (C) damped a_f of neighboring positions $x^{(2)}$ and $x^{(4)}$, where there were damping controls with opposite phases for the second and fourth electrodes.

We first focus on Figure 4, where the target electrode is the third electrode. The frequency f_s of sound stimuli was set to the resonant frequency at the position of the third electrode, $x^{(3)} = 13$ mm (i.e., $f_s = f_{res}^{(3)} = 6.25$ kHz). The result of protocol (A) shows that the amplitude was 4.6 nm at the electrode position of $x^{(3)} = 13$ mm, which was slightly smaller than the amplitude $a_f = 5.0$ nm at $x = 14$ mm, even though the sound stimuli were set to the resonant frequency at $x = 13$ mm. This is attributed to the fact that the width in the y-direction, $l(x)$, was larger for $x = 14$ mm. Note that the amplitude tended to be larger if the width $l(x)$ was wider. Moreover, we observed local maxima at $x = 21$ mm and $x = 27$ mm. These local maxima were related to the oscillation mode in the x direction, which can also be observed in Figure 2 with $f_s = 6.6$ kHz and in Reference [2].

At this stage, the amplitude at the resonant position was not prominent. We then tried protocol (B), where the electrical stimuli of $f_e^{(3)} = f_{res}^{(3)} = 6.25$ kHz were applied in addition to the sound stimuli. We observed an obvious amplification as a result of the electrical stimuli, and the amplitude at $x^{(3)}$ was 8.9 nm. We finally tried protocol (C), applying the electrical stimuli of $f_e^{(2)} = f_e^{(4)} = 6.25$ kHz with $\phi^{(2)} = \phi_{min}^{(2)}$ and $\phi^{(4)} = \phi_{min}^{(4)}$. Figure 4b shows that the amplitudes at $x^{(2)}$ and $x^{(4)}$ were successfully suppressed.

Figure 5 with the target electrode being the fourth electrode shows a situation similar to that of Figure 4 with the target electrode being the third electrode. The resonant frequency was $f_{res}^{(4)} = 5.10$ kHz. For protocol (A), the amplitude at $x^{(4)} = 17$ mm was at a maximum (4.0 nm), although we saw smaller peaks at $x = 20.5$ mm and $x = 25$ mm. As in the case of Figure 4, these smaller peaks must be due to the higher vibration mode at larger x, as also seen in Figure 2 with $f_s = 5.6$ kHz. With the electrical stimuli at the fourth electrode in protocol (B), the amplitude at $x^{(4)}$ was amplified to 7.9 nm. However, the small peaks at $x = 20.5$ mm and $x = 25$ mm were also amplified. For protocol (C), where the sinusoidal electrical stimuli with opposite phase $\phi^{(i_t \pm 1)} = \phi_{min}^{(i_t \pm 1)}$ were applied to the third and fifth electrodes, the amplitudes at $x^{(3)}$ and $x^{(5)}$ were lower, but the amplitude at $x^{(4)}$ also decreased from 7.9 to 6.7 nm.

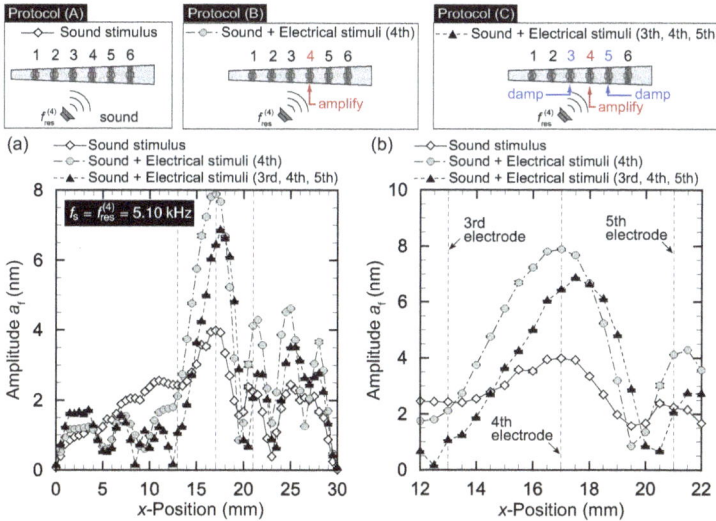

Figure 5. Vibration amplitude distribution for protocol (A): only sinusoidal sound stimuli with $f_s = 5.10$ kHz; protocol (B): sound and electrical stimuli applied to the fourth electrode; and protocol (C): sound and electrical stimuli applied to the third to fifth electrodes. (**a**) Entire view and (**b**) magnified version near $x^{(4)}$. Protocol (B) succeeded in amplifying a_f at resonant position $x^{(4)} = 17$ mm. In addition to the amplification in Protocol (B), Protocol (C) suppressed a_f of neighboring positions $x^{(3)}$ and $x^{(5)}$, where there were damping controls with opposite phases for the third and fifth electrodes.

The case where the target electrode was the fifth electrode is described in Figure 6, which should be compared with Figures 4 and 5 for the target electrode being the third and fourth electrodes, respectively. The resonant frequency was $f_{res}^{(5)} = 4.66$ kHz. The amplitude at $x^{(5)}$ was 2.3 nm in protocol (A). As in the cases of $i_t = 3$ and $i_t = 4$, we observed a smaller peak at $x = 27$ mm. The electrical stimuli of the fifth electrode in protocol (B) led to the amplification of amplitude at $x^{(5)}$, and $a_f = 4.8$ nm was obtained. However, the smaller peak at $x = 27$ mm was also amplified. In protocol (C), where damping control was carried out for the fourth and sixth electrodes, the amplitudes at $x^{(4)}$ and $x^{(6)}$ were lower as shown in Figure 6b.

To quantify the effect of vibration control using electrical stimuli, we define a parameter $S^{(i)}$ as

$$S^{(i)} = \frac{a_{max}^{(i)}}{x_L^{(i)} - x_R^{(i)}},$$

(2)

where $a_{max}^{(i)}$ is the maximum amplitude for all x, and $x_L^{(i)}$ and $x_R^{(i)}$ are positions such that the amplitude a_f becomes half of $a_{max}^{(i)}$. If $S^{(i)}$ is large, the amplitude at the resonant electrode is high and localized; that is, the response of the device is improved and $S^{(i)}$ can be considered as a variant of Q factors. The response factors $S^{(i_t)}$ ($i_t = 3, 4$, and 5) for the above protocols are summarized in Table 1. For $i_t = 3$ and 5, a comparison of protocols (A) and (B) shows that $S^{(i_t)}$ in protocol (B) became 2.2 times that in protocol (A). This is because we set $V_e^{(i_t)}$ such that the amplitude doubled in amplifying control. For $i_t = 4$, the increase in $S^{(4)}$ in protocol (B) was prominent, and the ratio between $S^{(4)}$ in protocols (B) and (A) took the value 1.67/0.35 = 4.77. This is because the peak near the maximum amplitude was not sharp when only sound was applied (protocol (A)), as shown in Figure 5 . We therefore conclude that vibration control worked well for better frequency selectivity, especially when the spatial response of the artificial cochlear epithelium was not sharp. The comparison between the response factor $S^{(i_t)}$ for protocols (B) and (C) showed that additional damping control increased $S^{(i_t)}$ for $i_t = 3$ and 5, but slightly reduced for $i_t = 4$. This decrease was due to the suppressed maximum amplitude in protocol (C) for $i_t = 4$. To increase the $S^{(i_t)}$ value, it is necessary to predict the motion of the trapezoidal membrane induced by localized electrical stimuli.

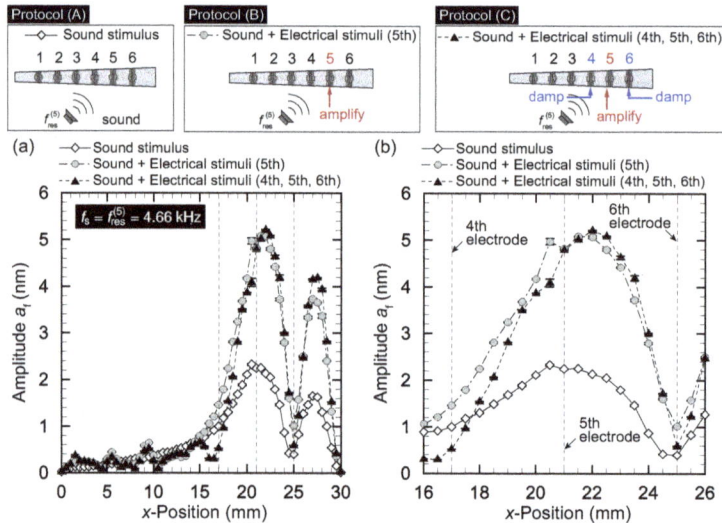

Figure 6. Vibration amplitude distribution for protocol (A): only sinusoidal sound stimuli with $f_s = 4.66$ kHz; protocol (B): sound and electrical stimuli applied to the fifth electrode; and protocol (C): sound and electrical stimuli applied to the fourth to sixth electrodes. (**a**) Entire view and (**b**) magnified version near $x^{(5)}$. Protocol (B) succeeded in amplifying a_f at resonant position $x^{(5)} = 21$ mm. In addition to the amplification in Protocol (B), Protocol (C) damped a_f of neighboring positions $x^{(4)}$ and $x^{(6)}$, where there were damping controls with opposite phases for the fourth and sixth electrodes.

Table 1. Response factor $S^{(i_t)}$ for protocols (A), (B), and (C) with $i_t = 3, 4,$ and 5.

	$S^{(i_t)} \times 10^6$		
	Protocol (A)	Protocol (B)	Protocol (C)
Sound Stimuli	On	On	On
Electrical Stimuli Applied to the i_t-th Electrode for Amplification Control	Off	On	On
Electrical Stimuli Applied to the $(i_t \pm 1)$-th Electrodes for Damping Control	Off	Off	On
$i_t = 3$	0.90	2.00	2.42
$i_t = 4$	0.35	1.67	1.53
$i_t = 5$	0.42	0.91	0.99

The sound pressure level P_s in the above experiments is now described. We chose the magnitude of sound stimuli $\bar{V}_s$ such that the amplitude a_f was on the order of 1 nm, because the amplitude of the vibration of the basilar membrane typically falls in this range. For $i_t = 3, 4,$ and 5, P_s were obtained as 92, 69, and 93 dB SPL. These P_s are within the the wide dynamic range of human hearing from 0 to 120 dB SPL.

3.2.3. Nonlinear Response of the Device to Sound

Human hearing has a dynamic range from 0 to 120 dB in amplitude, although the range of displacement of the cochlear epithelium is from 0.1 to 10 nm. Therefore, the magnitude of the pressure disturbance over 10^6 times (from 20 µPa to 20 Pa) is compressed to the amplitude range over 100 times (0.1 nm to 10 nm). Such a compression is caused by the outer hair cells in the organ of Corti, which elongate and shorten according to sound stimuli to control the movement of basilar membrane [33]. In Reference [34], using a cantilever device, a nonlinear feedback control which realizes the same compression rate as the human hearing is achieved by introducing a cubic damping term with respect to the cantilever velocity in the control signal. Here, we demonstrate that the same compression rate can be achieved using our MEMS-fabricated artificial cochlea.

Without any electrical stimuli, our device showed a linear response to the sound; that is, $a_f \sim p$, where p is the magnitude of the pressure disturbance induced by sound. Figure 7 shows the amplitude at the position of the fifth electrode when a sound stimuli with resonant frequency at this position, $f_s = f_{res}^{(5)} = 4.85$ kHz, was applied. It is seen that, without electrical stimuli, the amplitude showed a linear response to the sound pressure level. Note that the sound pressure level is related to p as $P_s = 20 \log_{10}(p/p_0)$, where $p_0 = 20$ µPa is the lowest-level of human hearing. To realize the nonlinear response, we applied an electrical stimuli to the fifth electrode, where the magnitude of the electrical signal $\bar{V}_e^{(5)}$ and the phase $\phi^{(5)}$ are shown in the inset. As shown in Figure 7, the amplitude with electrical stimuli was amplified for $P_s \leq 90$ dB SPL and was damped for $P_s \geq 90$ dB SPL, realizing the nonlinear response. The lines are the fitting curves to the experimental results shown by symbols. We used $a_f \sim p$ and $a_f \sim p^{1/3}$ as fitting curves for the results without electrical stimuli and those with electrical stimuli, respectively. This simple demonstration shows that the present concept of mimicking outer hair cells is feasible within the range of sound pressure level investigated. However, as shown in the inset of Figure 7, electrical signal with a couple of volts are necessary to dampen large vibration. Therefore, the device performance is limited by the electrical power supply.

3.3. Recognition of Vibration Using Electrical Outputs

Owing to the vibration of the membrane, the i-th recognition electrodes generate an electrical output with amplitude $\bar{V}_{rec}^{(i)}$. We measured $\bar{V}_{rec}^{(i)}$ and investigated their relation to the vibration amplitude $a_f^{(i)}$ of the membrane and the magnitude of the sound stimuli $\bar{V}_s$ shown in Figure 1c.

Note that $\bar{V}_{\text{rec}}^{(i)}$ is proportional to the magnitude of the strain of the piezoelectric membrane, which is closely related to the vibration amplitude $a_f^{(i)}$.

We present results for two cases, $f_s = f_{\text{res}}^{(4)} = 5.84$ kHz and $f_s = f_{\text{res}}^{(5)} = 4.56$ kHz, and analyze $\bar{V}_{\text{rec}}^{(i)}$ for $i = 1, 2, \cdots, 6$. The results for these two frequencies are respectively shown in Figures 8 and 9. Figure 8a shows the relationship between amplitude a_f at $x^{(i)}$ ($i = 1, 2, \cdots, 6$) measured by the LDV and sound pressure level P_s for $f_s = f_{\text{res}}^{(4)} = 5.84$ kHz. It is seen that the amplitude had a linear relationship with P_s, as expected. Figure 8b presents the relationship between electrical output $\bar{V}_{\text{rec}}^{(i)}$ ($i = 1, 2, \cdots, 6$) and amplitude a_f at $x^{(i)}$ for $f_s = f_{\text{res}}^{(4)} = 5.84$ kHz. $\bar{V}_{\text{rec}}^{(i)}$ is linearly correlated with $a_f^{(i)}$. However, the coefficients of proportionality depend on the electrode; that is, when we define a displacement–output conversion factor $\gamma^{(i)} = \bar{V}_{\text{rec}}^{(i)}/a_f^{(i)}$ (V/m), $\gamma^{(i)}$ changes according to i. The conversion factor $\gamma^{(i)}$ tends to be smaller for larger $x^{(i)}$. This is because the same amplitude results in larger deformation of the membrane when $l(x)$ is smaller. In Section 3.4, we use $V_{\text{rec}}^{(i)}$ to recognize the electrode with maximum amplitude. The values of $\gamma^{(i)}$ are necessary for correct displacement–output conversion and feedback control.

Figure 9a confirms the linear relationship between amplitude $a_f^{(i)}$ and P_s for $f_s = f_{\text{res}}^{(5)} = 4.56$ kHz; that is, the magnitude of the membrane vibration is proportional to that of sound stimuli. Figure 9b clearly shows that $\bar{V}_{\text{rec}}^{(i)} = \gamma^{(i)} a_f^{(i)}$, but the values of $\gamma^{(i)}$ are different from those in Figure 8b. For $f_s = f_{\text{res}}^{(5)} = 4.56$ kHz, the conversion factor $\gamma^{(i)}$ tends to be smaller for larger $x^{(i)}$, as in the case of $f_s = f_{\text{res}}^{(4)} = 5.84$ kHz. This indicates that the amplitude $a_f^{(i)}$ and $\bar{V}_{\text{rec}}^{(i)}$ have a non-trivial relationship. Further studies on the dimensions of the electrode and strain field are needed to clarify this relationship. This topic is left as future work, because the main goal of the present paper is to show the feasibility of vibration control using electrical stimuli.

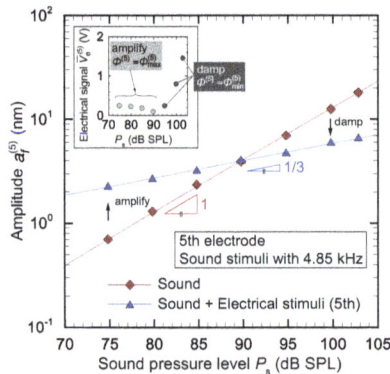

Figure 7. Relationship between amplitude a_f at $x^{(5)}$ and sound pressure level P_s for $f_s = 4.85$ kHz. The case with sound stimuli showed a linear response, while the case with both sound and electrical stimuli showed a nonlinear response with a power $1/3$. The electrical signal $\bar{V}_e^{(5)}$ used in the experiment is plotted in the inset.

3.4. Mimicking the Function of Outer Hair Cells by Electrical Feedback Control

This section describes the results of feedback control of vibration by applying electrical stimuli to the control electrodes and using the electrical output from the recognition electrodes, which are presented in Figure 1a. A schematic diagram of the experiment is described in Figure 10a. The parameters of the electrical stimuli, such as $\phi^{(i)}$, were determined before the experiment in the same manner as in Section 3.2.2. As described in Section 3.3, it was necessary to determine $\gamma^{(i)}$ to relate the output voltage $\bar{V}_{rec}^{(i)}$ from the recognition electrodes and the amplitude $a_f^{(i)}$. These values of $\gamma^{(i)}$ for each electrode and for each frequency were obtained prior to the following experiment.

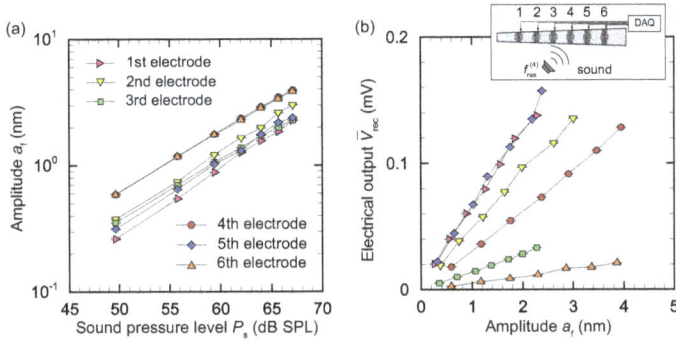

Figure 8. (**a**) Relationship between amplitude a_f at $x^{(i)}$ ($i = 1, 2, \cdots, 6$) and sound pressure level P_s for $f_s = f_{res}^{(4)} = 5.84$ kHz. The membrane vibration increased with the sound pressure level applied to the artificial cochlear epithelium. (**b**) Relationship between electrical output $\bar{V}_{rec}^{(i)}$ ($i = 1, 2, \cdots, 6$) and amplitude a_f at $x^{(i)}$ for $f_s = f_{res}^{(4)} = 5.84$ kHz. Amplitude a_f and electrical output $\bar{V}_{rec}$ from the recognition electrodes had a linear relationship.

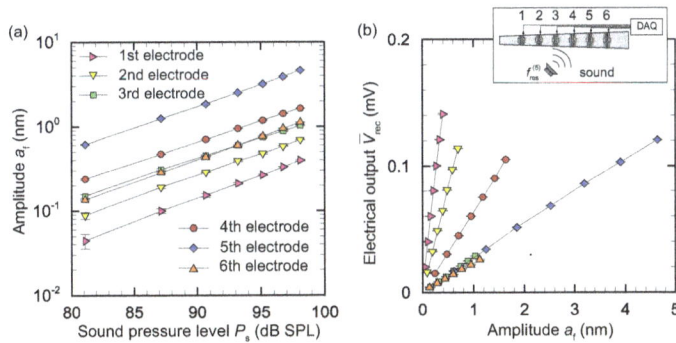

Figure 9. (**a**) Relationship between amplitude a_f at $x^{(i)}$ ($i = 1, 2, \cdots, 6$) and sound pressure level P_s for $f_s = f_{res}^{(5)} = 4.56$ kHz. The membrane vibration increased with the sound pressure level applied to the artificial cochlear epithelium; (**b**) Relationship between electrical output $\bar{V}_{rec}^{(i)}$ ($i = 1, 2, \cdots, 6$) and amplitude a_f at $x^{(i)}$ for $f_s = f_{res}^{(5)} = 4.56$ kHz. Amplitude a_f and electrical output $\bar{V}_{rec}$ from the recognition electrodes had a linear relationship.

The experiment comprised recognition and control stages. We carried out two cycles of these stages sequentially, as shown in Figure 10a. In the first and second cycles, sound stimuli with $f_s = f_{res}^{(5)}$ and $f_s = f_{res}^{(4)}$ were applied, respectively. The frequency of sound was changed between the first and second cycles to demonstrate that the present system responded to the frequency change. As a comprehensible demonstration, we chose $f_{res}^{(5)}$ for the first cycle and $f_{res}^{(4)}$ for the second cycle, but other frequencies can be chosen. In the recognition stage, no electrical stimuli were applied and electrical measurements of $\bar{V}_{rec}^{(i)}$ were automatically made to find the resonant position. In the present demonstration, these resonant positions for first and second cycles were respectively $x^{(5)}$ and $x^{(4)}$. Note that the LDV measurements were also made to confirm that the amplitude was magnified/damped as expected, but were not used for feedback control. Each stage took tens of seconds because the x–y auto-stage moved at a speed of 7 mm·s^{-1} to the measurement position (i.e., $x^{(i)}$, $i = 1, 2, \cdots, 6$) and waited five seconds before the measurement of vibration. The purpose of the present study was to develop and confirm a prototype principle of mimicking the outer hair cell as a first step. This waiting time was necessary for precise measurement because any tiny oscillation may affect the result. In the control stage, electrical stimuli were applied to amplify the vibration of the resonant position found in the recognition stage. As in protocol (C) described in Section 3.2.2, damping control was also carried out for neighboring electrodes. The results of the first and second cycles are respectively shown in Figure 10b,c. In the recognition stage of the first cycle, the position of the fifth electrode was detected as the resonant position as seen from the values of the modified electrical output $\tilde{V}_{rec}^{(i)}/\gamma^{(i)}(=a_f^{(i)})$. The electrical stimuli were then applied in the control stage of the first cycle to yield a prominent peak at the fifth position. Between the first and second cycles ($t \approx 50$ s), the frequency of sound changed to $f_{res}^{(4)}$. The measurement of electrical outputs in the recognition stage of the second cycle yielded that $\tilde{V}_{rec}^{(4)}/\gamma^{(4)}$ was the maximum and the position of the fourth electrode was thus detected as a resonant position. In the control stage of the second cycle, we successfully amplified the vibration at the fourth electrode while suppressing amplification at the neighboring electrodes (third and fifth), although the fifth electrode was slightly amplified.

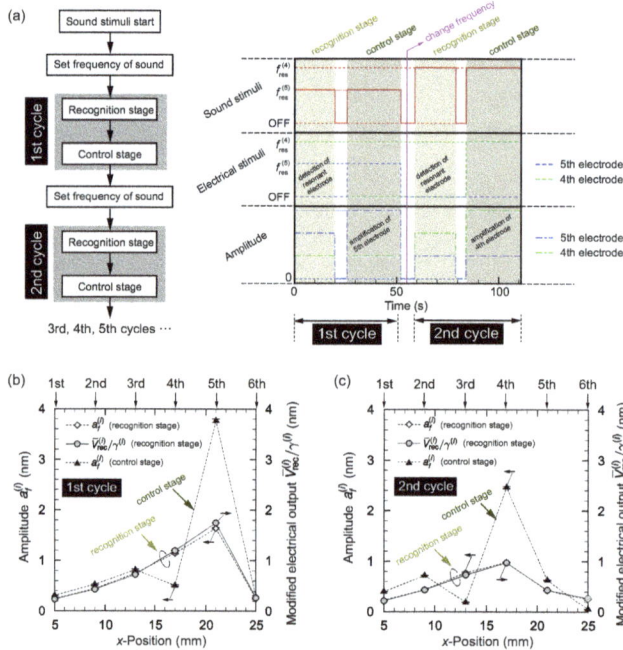

Figure 10. (a) Schematic diagram of the feedback control experiment. The experiments comprised recognition and control stages. We carried out two cycles of these stages sequentially as shown in Figure 10a. In the first and second cycles, sound stimuli with $f_s = f_{res}^{(5)}$ and $f_s = f_{res}^{(4)}$ were applied, respectively. The frequency of sound was changed between the first and second cycles to demonstrate that the present system responds to the frequency change. In the recognition stage, the electrical output $\tilde{V}_{rec}^{(i)}$ was automatically measured to find the resonant position. In the control stage, the electrical stimuli were applied to amplify the vibration of the resonant position found in the recognition stage. Amplitude distribution in (b) the first cycle and (c) the second cycle. In the first cycle shown in panel (b), the fifth electrode was detected as the resonant position in the recognition stage, and the amplitude of vibration was amplified only near the fifth electrode in the control stage. In the second cycle shown in panel (c), the fourth electrode was detected as the resonant position in the recognition stage, and the amplitude of vibration was amplified only near the fourth electrode in the control stage.

We define another response factor $Q^{(i)}$ as

$$Q^{(i)} = \frac{a_f^{(i)}}{(a_f^{(i+1)} + a_f^{(i-1)})/2}. \tag{3}$$

In the above four experiments, for the first/second cycles and recognition/control states, $Q^{(i)}$ values were obtained as given in Table 2. The table shows that the $Q^{(i)}$ value was higher in the control stage and the magnification ratios were 4.03 and 3.64 for the first and second cycles, respectively. These results indicate that the present experimental system was capable of increasing the performance of frequency selectivity of the artificial cochlear epithelium by mimicking the function of outer hair cells.

Table 2. Response factor $Q^{(i)}$ for recognition/control stages of the first/second cycles presented in Figure 10.

	$Q^{(i)}$		$\dfrac{Q^{(i)} \text{ of Recognition Stage}}{Q^{(i)} \text{ of Control Stage}}$
	Recognition Stage	**Control Stage**	
First Cycle ($i_t = 5$)	2.35 ± 0.06	9.48 ± 0.49	4.03
Second Cycle ($i_t = 4$)	1.64 ± 0.03	5.96 ± 0.15	3.64

4. Conclusions

We propose an artificial cochlear epithelium which mimics the function of an outer hair cell using feedback electrical stimuli. The main outcomes of the present paper are summarized as follows.

1. On the basis of a previous device [2], we developed a new design of an artificial cochlear epithelium with recognition and control electrodes. These electrodes are used to mimic the functions of the basilar membrane, inner hair cells, and outer hair cells.
2. Recognition of the resonant position and control of the vibration amplitude at the resonant position are realized using the electrode pattern of the present device. The method uses the local electrical stimuli through patterned electrodes fabricated on a PVDF membrane with a trapezoidal support. Parameters of the electrical stimuli were experimentally determined for each electrode to improve the response of the artificial cochlear epithelium.
3. A demonstration of the feedback control of membrane vibration was carried out by alternating the frequency of sound stimuli during a single run of the experiment. The present device automatically responds to a change in the sound frequency and amplifies the vibration amplitude at the resonant position.

There are ways to improve the present control method. Firstly, it is important to design the recognition electrodes to evaluate the amplitude of membrane vibration quantitatively. That is to say, one needs to control values of the displacement–output conversion factor of the i-th electrode (i.e., $\gamma^{(i)}$ in Section 3.3) by changing the dimensions of the recognition electrodes. As an alternative solution, one may use a machine-learning technique to search for appropriate control parameters of electrical stimuli (e.g., $\phi^{(i)}$) and to construct a database of $\gamma^{(i)}$. Another important direction of improvement is minimization of the experimental setup. We constructed an automation system for the present study, but the system obviously cannot be integrated with an actual artificial cochlea. It is necessary to develop an equivalent circuit system using micro-fabrication technologies for further investigation, such as animal tests. To sustain electrical power to activate devices, one may consider using an endocochlear potential maintained in the cochlea [37,38], which has also been proposed as a biological battery [39]. The cochlear shape is important in low-frequency hearing [40,41], and thus evaluation of the device in an environment similar to that of the cochlea is also necessary for the optimal design of wide-range frequency selectivity.

Author Contributions: T.T. and S.K. conceived and designed the research; T.T, A.N., and H.Y. constructed the measurement system and performed the experiments; T.T. and A.N. analyzed the data; T.T. and S.K. wrote the paper.

Acknowledgments: This research is supported by the Japan Agency for Medical Research and Development under Grant Number JP17gm0810004.

Conflicts of Interest: The authors declare no conflict of interest.

Appendix A. Fabrication of the PVDF Membrane with Patterned Electrodes

The fabrication method of the patterned electrode on the PVDF membrane is described as follows· and summarized in Figure A1. An Al thin film with thickness of 50–60 nm was deposited on both sides of a PVDF membrane with thickness of 40 µm (KF piezo film, Kureha, Tokyo, Japan). The PVDF membrane was attached to a glass substrate for the following photolithography process. A positive photoresist (AZ5214E, Merck, Darmstadt, Germany) was spin-coated on the PVDF membrane at 3000 rpm. After a prebake for 10 min at 80 degrees, the membrane was exposed to ultraviolet light at 6 mJ/cm^2 through a photomask. To expose the membrane uniformly, three glass diffuser plates were placed between the light source and membrane. The exposed photoresist was developed by immersing the membrane in developer solution (AZ300MIF, Merck, Darmstadt, Germany) for 8 min. During the development process, the Al film except for the area covered by the cured photoresist was etched, and the PVDF membrane with a patterned electrode was thus obtained as shown in Figure 1b. Finally, the membrane was immersed in ethanol for 2 min to remove the photoresist. The Al thin film on the backside of the PVDF membrane is protected and maintained by the glass substrate.

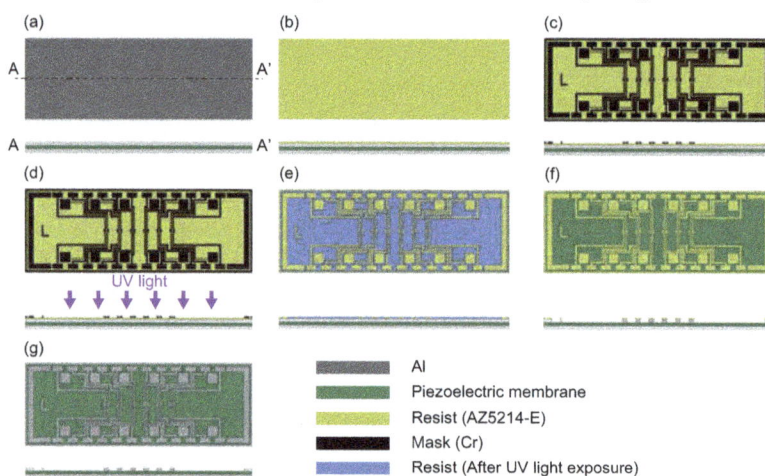

Figure A1. Fabrication process of the PVDF membrane with patterned electrodes. (**a**) Al-deposited piezoelectric membrane; (**b**) Spincoating of photoresist AZ5214-E; (**c**) Covering the photomask; (**d**) Exposure to ultraviolet light; (**e**) Removing the photomask; (**f**) Development and etching; (**g**) Lifting off.

References

1. Inaoka, T.; Shintaku, H.; Nakagawa, T.; Kawano, S.; Ogita, H.; Sakamoto, T.; Hamanishi, S.; Wada, H.; Ito, J. Piezoelectric materials mimic the function of the cochlear sensory epithelium. *Proc. Natl. Acad. Sci. USA* **2011**, *108*, 18390–18395. [CrossRef] [PubMed]
2. Shintaku, H.; Nakagawa, T.; Kitagawa, D.; Tanujaya, H.; Kawano, S.; Ito, J. Development of piezoelectric acoustic sensor with frequency selectivity for artificial cochlea. *Sens. Actuators A Phys.* **2010**, *158*, 183–192. [CrossRef]

3. Kawano, S.; Ito, J.; Nakagawa, T.; Shintaku, H. Artificial Sensory Epithelium. U.S. Patent 9,566,428, 14 February 2017.

4. Von Békésy, G.; Wever, E.G. *Experiments in Hearing*; McGraw-Hill: New York, NY, USA, 1960; Volume 8.

5. Warchol, M.E.; Lambert, P.R.; Goldstein, B.J.; Forge, A.; Corwin, J.T. Regenerative proliferation in inner ear sensory epithelia from adult guinea pigs and humans. *Science* **1993**, *259*, 1619–1622. [CrossRef] [PubMed]

6. Clark, G.M.; Tong, Y.; Black, R.; Forster, I.; Patrick, J.; Dewhurst, D. A multiple electrode cochlear implant. *J. Laryngol. Otol.* **1977**, *91*, 935–945. [CrossRef] [PubMed]

7. Zeng, F.G. Trends in cochlear implants. *Trends Amplif.* **2004**, *8*, 1–34. [CrossRef] [PubMed]

8. Shintaku, H.; Tateno, T.; Tsuchioka, N.; Tanujaya, H.; Nakagawa, T.; Ito, J.; Kawano, S. Culturing neurons on MEMS fabricated P (VDF-TrFE) films for implantable artificial cochlea. *J. Biomech. Sci. Eng.* **2010**, *5*, 229–235. [CrossRef]

9. Shintaku, H.; Kobayashi, T.; Zusho, K.; Kotera, H.; Kawano, S. Wide-range frequency selectivity in an acoustic sensor fabricated using a microbeam array with non-uniform thickness. *J. Micromech. Microeng.* **2013**, *23*, 115014. [CrossRef]

10. Shintaku, H.; Inaoka, T.; Nakagawa, T.; Kawano, S.; Ito, J. Electrically evoked auditory brainstem response by using bionic auditory membrane in guinea pigs. *J. Biomech. Sci. Eng.* **2013**, *8*, 198–208. [CrossRef]

11. Kim, S.; Song, W.J.; Jang, J.; Jang, J.H.; Choi, H. Mechanical frequency selectivity of an artificial basilar membrane using a beam array with narrow supports. *J. Micromech. Microeng.* **2013**, *23*, 095018. [CrossRef]

12. Tateno, T.; Nishikawa, J.; Tsuchioka, N.; Shintaku, H.; Kawano, S. A hardware model of the auditory periphery to transduce acoustic signals into neural activity. *Front. Neuroeng.* **2013**, *6*, 12. [CrossRef] [PubMed]

13. Jang, J.; Kim, S.; Sly, D.J.; O'fleary, S.J.; Choi, H. MEMS piezoelectric artificial basilar membrane with passive frequency selectivity for short pulse width signal modulation. *Sens. Actuators A Phys.* **2013**, *203*, 6–10. [CrossRef]

14. Tanujaya, H.; Shintaku, H.; Kitagawa, D.; Adianto, A.; Susilodinata, S.; Kawano, S. Experimental and analytical study approach of Artificial basilar membrane prototype (ABMP). *J. Eng. Technol. Sci.* **2013**, *45*, 61–72. [CrossRef]

15. Tanujaya, H.; Kawano, S. Experimental study of vibration of prototype auditory membrane. *Appl. Mech. Mater.* **2014**, *493*, 372–377. [CrossRef]

16. Lee, H.S.; Chung, J.; Hwang, G.T.; Jeong, C.K.; Jung, Y.; Kwak, J.H.; Kang, H.; Byun, M.; Kim, W.D.; Hur, S.; et al. Flexible inorganic piezoelectric acoustic nanosensors for biomimetic artificial hair cells. *Adv. Funct. Mater.* **2014**, *24*, 6914–6921. [CrossRef]

17. Jang, J.; Lee, J.; Woo, S.; Sly, D.J.; Campbell, L.J.; Cho, J.H.; O'fLeary, S.J.; Park, M.H.; Han, S.; Choi, J.W.; et al. A microelectromechanical system artificial basilar membrane based on a piezoelectric cantilever array and its characterization using an animal model. *Sci. Rep.* **2015**, *5*, 12447. [CrossRef] [PubMed]

18. Jang, J.; Jang, J.H.; Choi, H. Biomimetic artificial basilar membranes for next-generation cochlear implants. *Adv. Healthc. Mater.* **2017**, *6*, 1700674. [CrossRef] [PubMed]

19. Jang, J.; Jang, J.H.; Choi, H. MEMS flexible artificial basilar membrane fabricated from piezoelectric aluminum nitride on an SU-8 substrate. *J. Micromech. Microeng.* **2017**, *27*, 075006. [CrossRef]

20. Dagdeviren, C.; Yang, B.D.; Su, Y.; Tran, P.L.; Joe, P.; Anderson, E.; Xia, J.; Doraiswamy, V.; Dehdashti, B.; Feng, X.; et al. Conformal piezoelectric energy harvesting and storage from motions of the heart, lung, and diaphragm. *Proc. Natl. Acad. Sci. USA* **2014**, *111*, 1927–1932. [CrossRef] [PubMed]

21. Asadnia, M.; Kottapalli, A.G.P.; Karavitaki, K.D.; Warkiani, M.E.; Miao, J.; Corey, D.P.; Triantafyllou, M. From biological cilia to artificial flow sensors: Biomimetic soft polymer nanosensors with high sensing performance. *Sci. Rep.* **2016**, *6*, 32955. [CrossRef] [PubMed]

22. Ko, S.C.; Kim, Y.C.; Lee, S.S.; Choi, S.H.; Kim, S.R. Micromachined piezoelectric membrane acoustic device. *Sens. Actuators A Phys.* **2003**, *103*, 130–134. [CrossRef]

23. Luo, Y.; Gan, R.; Wan, S.; Xu, R.; Zhou, H. Design and analysis of a MEMS-based bifurcate-shape piezoelectric energy harvester. *AIP Adv.* **2016**, *6*, 045319. [CrossRef]

24. Tona, Y.; Inaoka, T.; Ito, J.; Kawano, S.; Nakagawa, T. Development of an electrode for the artificial cochlear sensory epithelium. *Hear. Res.* **2015**, *330*, 106–112. [CrossRef] [PubMed]

25. Gold, T.; Pumphrey, R.J. Hearing. I. The cochlea as a frequency analyzer. *Proc. R. S. Lon. B Biol. Sci.* **1948**, *135*, 462–491. [CrossRef]

26. Gold, T. Hearing. II. The physical basis of the action of the cochlea. *Proc. R. S. Lond. B Biol. Sci.* **1948**, *135*, 492–498. [CrossRef]

27. Kemp, D.T. Evidence of mechanical nonlinearity and frequency selective wave amplification in the cochlea. *Arch. Oto Rhino Laryngol.* **1979**, *224*, 37–45. [CrossRef]

28. Rhode, W.S. Observations of the vibration of the basilar membrane in squirrel monkeys using the Mössbauer technique. *J. Acoust. Soc. Am.* **1971**, *49*, 1218–1231. [CrossRef]

29. Russell, I.; Nilsen, K. The location of the cochlear amplifier: spatial representation of a single tone on the guinea pig basilar membrane. *Proc. Natl. Acad. Sci. USA* **1997**, *94*, 2660–2664. [CrossRef] [PubMed]

30. Robles, L.; Ruggero, M.A. Mechanics of the mammalian cochlea. *Physiol. Rev.* **2001**, *81*, 1305–1352. [CrossRef] [PubMed]

31. Duke, T.; Jülicher, F. Active traveling wave in the cochlea. *Phys. Rev. Lett.* **2003**, *90*, 158101. [CrossRef] [PubMed]

32. Dong, W.; Olson, E.S. Detection of cochlear amplification and its activation. *Biophys. J.* **2013**, *105*, 1067–1078. [CrossRef] [PubMed]

33. Reichenbach, T.; Hudspeth, A. The physics of hearing: fluid mechanics and the active process of the inner ear. *Rep. Prog. Phys.* **2014**, *77*, 076601. [CrossRef] [PubMed]

34. Joyce, B.S.; Tarazaga, P.A. Mimicking the cochlear amplifier in a cantilever beam using nonlinear velocity feedback control. *Smart Mater. Struct.* **2014**, *23*, 075019. [CrossRef]

35. Shabana, A.A. *Theory of Vibration: Volume II: Discrete and Continuous Systems*; Springer Science & Business Media: Heidelberg/Berlin, Germany, 2012.

36. Merchant, S.N.; Nadol, J.B. *Schuknecht's Pathology of the Ear*; People's Medical Publishing House-USA: Shelton, CT, USA , 2010; p. 128.

37. Nin, F.; Hibino, H.; Murakami, S.; Suzuki, T.; Hisa, Y.; Kurachi, Y. Computational model of a circulation current that controls electrochemical properties in the mammalian cochlea. *Proc. Natl. Acad. Sci. USA* **2012**, *109*, 9191–9196. [CrossRef] [PubMed]

38. Adachi, N.; Yoshida, T.; Nin, F.; Ogata, G.; Yamaguchi, S.; Suzuki, T.; Komune, S.; Hisa, Y.; Hibino, H.; Kurachi, Y. The mechanism underlying maintenance of the endocochlear potential by the K+ transport system in fibrocytes of the inner ear. *J. Physiol.* **2013**, *591*, 4459–4472. [CrossRef] [PubMed]

39. Mercier, P.P.; Lysaght, A.C.; Bandyopadhyay, S.; Chandrakasan, A.P.; Stankovic, K.M. Energy extraction from the biologic battery in the inner ear. *Nat. Biotechnol.* **2012**, *30*, 1240–1243. [CrossRef] [PubMed]

40. Manoussaki, D.; Chadwick, R.S. Effects of geometry on fluid loading in a coiled cochlea. *SIAM J. Appl. Math.* **2000**, *61*, 369–386. [CrossRef]

41. Manoussaki, D.; Chadwick, R.S.; Ketten, D.R.; Arruda, J.; Dimitriadis, E.K.; O'Malley, J.T. The influence of cochlear shape on low-frequency hearing. *Proc. Natl. Acad. Sci. USA* **2008**, *105*, 6162–6166. [CrossRef] [PubMed]

micromachines

MDPI

Article

Miniaturization and High-Density Arrangement of Microcantilevers in Proximity and Tactile Sensor for Dexterous Gripping Control

Ryoma Araki [1], Takashi Abe [1], Haruo Noma [2] and Masayuki Sohgawa [1,*]

[1] Graduate School of Science and Technology, Niigata University, 8050 Ikarashi 2 no-cho, Nishi-ku, Niigata 950-2181, Japan; f16b078e@mail.cc.niigata-u.ac.jp (R.A.); memsabe@eng.niigata-u.ac.jp (T.A.)

[2] Ritsumeikan University, 1-1-1, Noji-higashi, Kusatsu, Shiga 525-8577, Japan; noma@media.ritsumei.ac.jp or hanoma@fc.ritsumei.ac.jp

* Correspondence: sohgawa@eng.niigata-u.ac.jp; Tel.: +81-25-262-7819

Received: 28 April 2018; Accepted: 12 June 2018; Published: 15 June 2018

Abstract: In this paper, in order to perform delicate and advanced grip control like human, a proximity and tactile combination sensor using miniaturized microcantilevers one-fifth the size of previous one as the detection part was newly developed. Microcantilevers were arranged with higher spatial density than in previous works and an interdigitated array electrode to enhance light sensitivity was added. It is found that the interdigitated array electrode can detect light with 1.6 times higher sensitivity than that in previous works and the newly fabricated microcantilevers have enough sensitivity to applied normal and shear loads. Therefore, more accurate detection of proximity distance and spatial distribution of contact force become available for dexterous gripping control to prevent 'overshooting', 'force control error', and 'slipping'.

Keywords: tactile sensor; proximity sensor; slipping detection; microcantilever

---

1. Introduction

In recent years, as the declining birthrate and aging population increase, the labor force declines, and the burden on nursing care increases in developed countries, including Japan [1,2]. On the other hand, by advances in automation technologies, robots are being introduced not only in the manufacturing industries but also in various fields such as agriculture and medical welfare, and it has received increasing attention [3–7]. By introducing robots to human tasks, it is expected to contribute reduction of personnel expenses, efficiency of work, and reduction of human burdens and risks [2,8]. However, there are a lot of problems in robotization. One of them is manipulation control such as gripping. Objects handled in the field of manufacturing industries are typically rigid and have stated shape with constant mass, thus there is hardly any obstacle to manipulation. However, in the case of considering objects with fragile body and indefinite and complex shape such as fruits and human body, precise manipulation control is necessary to no damage or no destruction during handling them [9]. Humans have dexterity enable to competently grip or handle objects with smart sensation, distinguishing the shape and hardness of objects. This is because human fingers are the most prominent part of discrimination ability as a tactile sensor by tactile receptors located high density, in addition owing to feedback and feedforward based on this information and proximity information by visual sense [10,11]. Therefore, even in robots, if prevention of 'overshooting', 'force control error', and 'slipping' is realized by acquisition of proximity information and contact information between a hand and target object using a sensor corresponding to proximity and tactile sense, dexterous gripping control similar to human capability is expected [12]. Although there are many studies on tactile sensors for accurate gripping control including the miniature sensor with microcantilevers embedded in polydimethylsiloxane (PDMS) [4,13], gripping control similar to a human needs not only tactile sensing but also sensing of contact information by the proximity sensor. Some studies of sensors that integrated proximity information with tactile information have been conducted in recent years. Mizoguchi et al. integrated on robot hand a tactile sensor using pressure-sensitive conductive rubber and a proximity sensor using optical elements [14]. On the other hand, Tsuji and Kohama reported a study on a proximity/tactile sensor based on change of static capacitance [15]. However, these sensors are relatively large and have complicated designs because they need assembly processes. Considering that the sensors are installed on a robot hand, space saving and high accuracy are valued traits. Furthermore, it is important that distributional contact information is detectable by arraying multiple sensors.

In our previous works, single element proximity and tactile combined sensor fabricated by micromachining process for manipulation control not only in the manufacturing industries but also in various fields has been developed [16]. In this sensor, normal and shear loads can be obtained distinctively using this sensor with three cantilevers embedded in PDMS elastomer by measurement of sensitivities to each axis load in advance. As compared to other devices which can detect both proximity and tactile information mentioned above, our sensor features a smaller and simpler design, and can be installed on manipulators of various shapes, thus promising high versatility and low cost due to mass production. In addition, combined proximity and tactile detection is implemented using a single small sensitive element without assembling; hence, no need for multiple systems, which makes space saving and easier arraying possible. We have also developed manipulation system using a miniature electromotive manipulator with this sensor has been constructed [17]. It has been shown that this manipulator system can grip objects without damage or destruction occurred by 'overshoot just after gripping', and 'force control error after gripping'. Furthermore, flexible objects with different hardnesses have been gripped by this system successfully. However, the detection part of this sensor is comparatively large at 290×200 µm in length and width, respectively, thus it is difficult to place the detection part with high density for detection similar to spatial acuity of tactile receptors (0.5 mm for Merkel cells [4]) of humans, and detection of normal and shear force distribution at micro scale have limitations. In addition, for proximity sensing, a LED separate from the sensor serving as a probe light source is needed [16], however, it brings increase of mounting area and the shadowing effect in detection just before contact. To decrease mounting area and prevent shadowing, a smaller LED chip with lower light intensity will be mounted on surface of the sensor, thus, light sensitivity of the sensor should be improved. In this work, in order to perform delicate and advanced grip control similar to human, cantilevers are miniaturized to locate at high density from previous one. Furthermore, interdigitated array electrodes which enhance light sensitivity to detect farther proximity distance have been integrated on the chip.

2. Design and Fabrication of Tactile Sensor

2.1. Design of the Sensor

Figure 1 shows a schematic diagram of cross-sectional view of the sensor. In our tactile sensor, a strain gauge is formed on the microcantilever embedded in PDMS as tactile detection part (right part of Figure 1). In previous works, the size of one cantilever is comparatively large as 290×200 µm, and it was difficult to locate the cantilevers with high density similar to tactile receptors of human. Therefore, in this work, we aim to (1) reduce the size of the cantilever; and (2) place the cantilevers more densely. Furthermore; (3) an interdigitated array electrode (left part of Figure 1) is newly designed to improve the sensitivity as proximity sensor [18]. The pattern of the new sensor was designed using IC layout CAD (LayoutEditor, juspertor GmbH, Unterhaching, Germany). Figure 2a,b shows design drawings of a conventional cantilever and a newly fabricated cantilever, respectively. Cr pattern and NiCr meander wiring are formed on the cantilever. Etching windows for sacrificial etching of SiO_2 under the cantilever are also formed around and in the pattern of the cantilever. The cantilever will be warped upward by residual stress in a Cr layer after sacrificial etching, as shown in Figure 1. Comparing Figure 2a,b, the area of the cantilever part is reduced to one-fifth from the conventional one, and it is possible to locate more densely. Figure 3 shows the interdigitated array electrode for light detection. When the sensor surface is irradiated with light, resistance, and depletion layer capacitance in the Si layer decrease because of generation of electron–hole pairs (photocarriers). Because the interdigitated array electrode is electrically connected to the Si layer via capacitance of Si_3N_4 insulation layer as AC circuit shown in Figure 1, the impedance of the electrode decreases with increase of the light intensity. Thus, the light intensity can be detected as decrease of the impedance of the electrode [16]. In previous works, the light is detected as impedance between wiring electrodes [16,17]. It is found that contribution of depletion layer capacitance to light sensitivity is larger than that of resistance in the Si layer [18], however, effect of resistance in the Si layer is comparably large because of gap between wiring electrodes (>100 μm). Therefore, to decrease the gaps between electrodes and effect of resistance, we employ interdigitated array electrodes with narrow gaps in this work. The size of the interdigitated electrode array is 500 × 500 μm, and it is located so as not to interfere with the cantilevers and wire. In addition, the interdigitated electrode array has mesh holes to increase the amount of incident light on Si. This is because it has been demonstrated that light sensitivity can be enhanced by forming a lot of mesh holes [18].

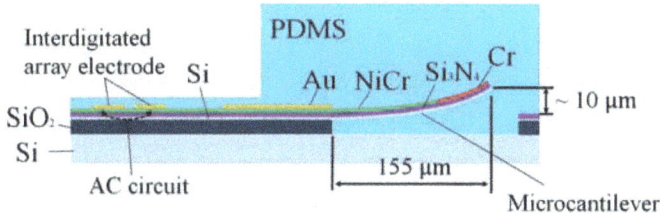

Figure 1. A schematic diagram of cross-sectional view of the sensor.

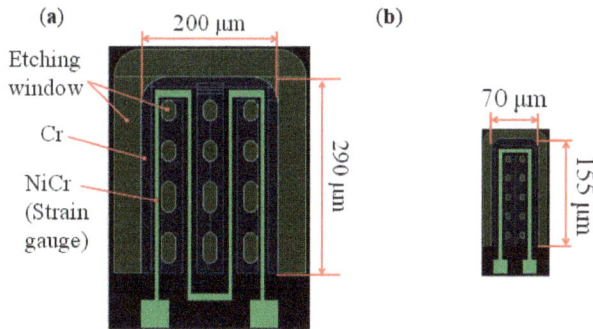

Figure 2. Design drawings of microcantilever in (**a**) previous works, and (**b**) this work.

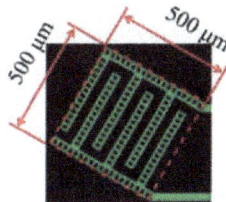

Figure 3. Design drawing of interdigitated array electrodes.

2.2. Fabrication of the Sensor

Figure 4a shows a cross-sectional view of fabrication process of the sensor, and Figure 4b shows a schematic illustration of cantilever. Si_3N_4 thin film was deposited as an insulating layer on a Si-on-insulator (SOI) wafer, then NiCr as thin film strain gauge layer, Au as a surface electrode and wiring layer, and Cr as a stress layer were deposited and patterned, respectively. Where, Si_3N_4, NiCr, and Au were deposited by sputtering method, and Cr was deposited by electron beam evaporation method. Thereafter, the Si active layer was anisotropically removed by reactive ion etching (RIE) and a pattern was formed for sacrificial layer etching. SiO_2 layer was etched in buffered hydrofluoric acid (BHF, 20%, 30 °C) for 150 min to release the upper structure as a cantilever. The released cantilever was warped by residual stress due to the difference in coefficient of thermal expansion between the Cr layer and the Si layer [19]. Although poly Si can be used as the layer for the cantilever structure, in this work, we employed single-crystal Si because of its uniformity of mechanical and optical characteristics. Furthermore, poly-dimethyl-siloxane (PDMS) elastomeric layer was coated on the chip in order to protect the chip surface, and a PDMS bump (1.6 mm diameter, 2 mm height) as a contact part was attached to center of the chip.

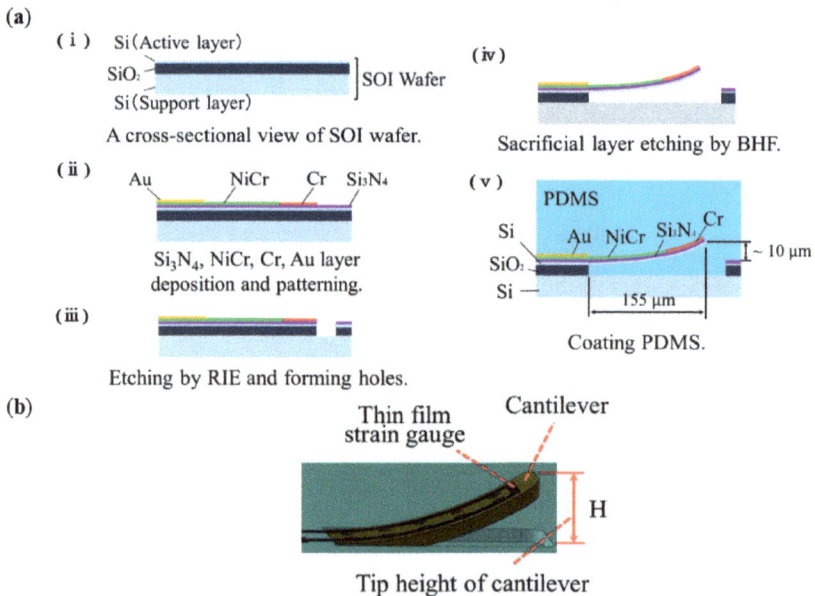

(a)

(i) Si (Active layer)
SiO_2 — SOI Wafer
Si (Support layer)

A cross-sectional view of SOI wafer.

(ii) Au   NiCr   Cr Si_3N_4

Si_3N_4, NiCr, Cr, Au layer deposition and patterning.

(iii)

Etching by RIE and forming holes.

(iv)

Sacrificial layer etching by BHF.

(v) PDMS
Si   Au NiCr Si_3N_4 Cr
SiO_2 ~ 10 µm
Si —   155 µm

Coating PDMS.

(b)

Thin film strain gauge   Cantilever

H

Tip height of cantilever

Figure 4. (**a**) Cross-sectional view of sensor fabrication method, and (**b**) schematic illustration of cantilever.

In this work, two types of sensors were newly fabricated. Figure 5a shows the sensor fabricated in previous works. On the other hand, Figure 5b,c shows a newly fabricated sensor named Type A and Type B, respectively. The size of the each sensor chip is 5×5 mm. In previous works, only three cantilevers could be located in the 1 mm diameter circle. However, in this work, owing to reduction in the size of the cantilever, it has become possible to locate 3 cantilevers in the 0.4 mm diameter circle (Type A) and 12 cantilevers at a higher density in the circle of 1 mm (Type B). In addition, by reducing the area occupied by the cantilever, it has also become possible to locate an interdigitated array electrode in the newly fabricated sensor. In the newly fabricated cantilever, the length and width are different from the previous one, thus the height of tip of warped cantilever is also different [20]. The sensitivities of the sensor to normal load and shear load depends on tip height of the cantilever.

Therefore, tip height of fabricated cantilever was measured with a laser displacement meter (LT-9000, Keyence, Osaka, Japan). Measurements were performed for a total of nine cantilevers in three sensors of Type A. As a result, it is confirmed that their average is 9.60 µm with the standard deviation of 0.70 µm which results in similar bending angle. This standard deviation value is smaller than that in previous works.

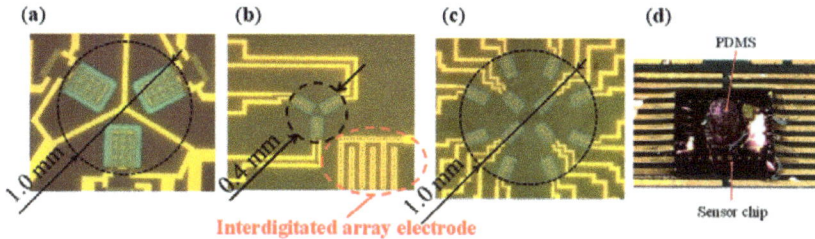

Figure 5. Optical microscopic images of (**a**) conventional sensor, (**b**) newly fabricated sensor of Type A, and (**c**) newly fabricated sensor of Type B, and (**d**) the sensor chip with PDMS.

3. Proximity and Tactile Measurement Principle

3.1. Optical Responsivity of Interdigitated Array Electrode

The tactile sensor fabricated in this work is employed single crystal silicon which is a typical semiconductor material as the main substrate material. Hence, when the sensor is irradiated with light, electron–hole pairs as photocarriers are generated by excitation of valence electrons due to the photoconductive effect, and electric resistivity and depletion layer capacitance in Si layer are modulated. By detecting these changes as an impedance change in the Si layer by reflected light from the object, proximity distance can be measured. To confirm the optical responsivity, the dependence of impedance change of the interdigitated array electrode on light illuminance was measured, as shown in Figure 6. This measurement was performed in a darkroom, and the impedance was measured using an LCR meter (Hioki 3532-50, Hioki E.E. Corporation, Nagano, Japan) at the measuring frequency of 5 MHz. The impedance decreases with increasing light illuminance in both cases of the sensor with and without interdigitated array electrode. It is found that the optical responsivity of the sensor with interdigitated array electrode is 1.6 times higher than that without interdigitated array electrode. This is because the change of depletion layer capacity when light is irradiated is increased by electrode of mesh structure. Therefore, proximity detection with higher sensitivity than previous works is expected using the sensor designed in this work.

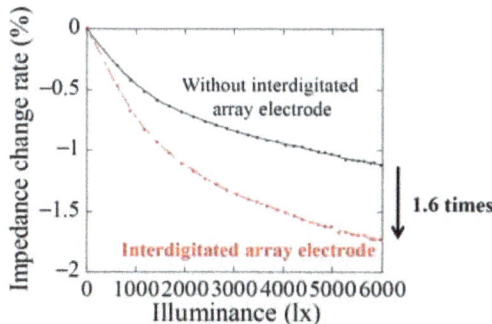

Figure 6. Optical responsivity of the sensor with and without interdigitated array electrodes, respectively.

3.2. Optical and Load Responsivity of Fabricated Cantilever

In order to detect the force and proximity separately, it is required that the strain gauge resistance on the cantilever is sensitive only to strain and has no sensitivity to light. Therefore, the illuminance dependency of the strain gauge resistance of the newly manufactured cantilever was measured. This measurement was performed in a darkroom, and the strain gauge resistance was measured by a digital multimeter (Advantest R6581, Advantest Corporation, Tokyo, Japan). Figure 7 shows resistance of the strain gauge as a function of light illuminance. It is found that the resistance is nearly independent of light illuminance. Thus, it is possible that the sensor can detect separately force and proximity.

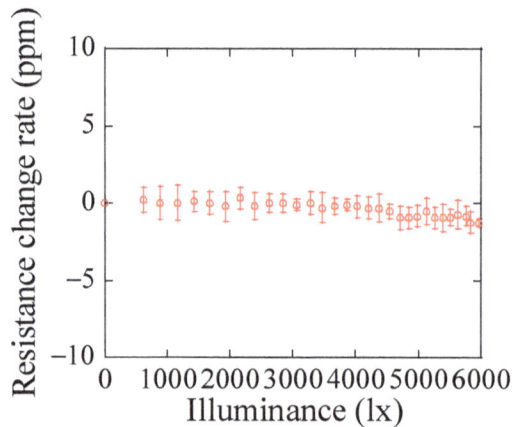

Figure 7. Optical responsivity of the strain gauge resistance.

Next, in order to measure the load response characteristics of the newly fabricated cantilever, resistance change of the strain gauge was measured when normal and shear loads were applied to the sensor. Figures 8 and 9 show the comparison of resistance changes for normal and shear loads between the sensors fabricated in previous and this works, respectively. From Figure 8, the newly fabricated cantilever has sensitivity to normal load, however, it is two-thirds lower than that of previous one. On the other hand, from Figure 9, it is found that sensitivity to shear load of the newly fabricated cantilever is approximately 2.1 times higher than that of the previous one. It is considered that this sensitivity difference is due to the size and tip height of the cantilever. The cantilever in previous cantilever has tip height of 30 μm and length of 290 μm, hence, its angle is approximately calculated as 5.9° using arc tangent. On the other hand, that of the new cantilever with tip height of 9.6 μm and length of 155 μm is calculated as 3.5°, which is smaller than previous one. Conversely, it is suggested that we can calibrate the sensitivities to normal and shear loads by controlling the tip height of the cantilever. In addition, it has been confirmed that the warp of the newly fabricated cantilever is not uniform and only its tip is locally lifted up. As a result, it is considered that the sensitivity is enhanced by the large deformation of the cantilever when applied shear load. From the above results, it is demonstrated that the sensor with miniaturized cantilever in this work has sensitivities to both normal and shear loads similar to that in previous works.

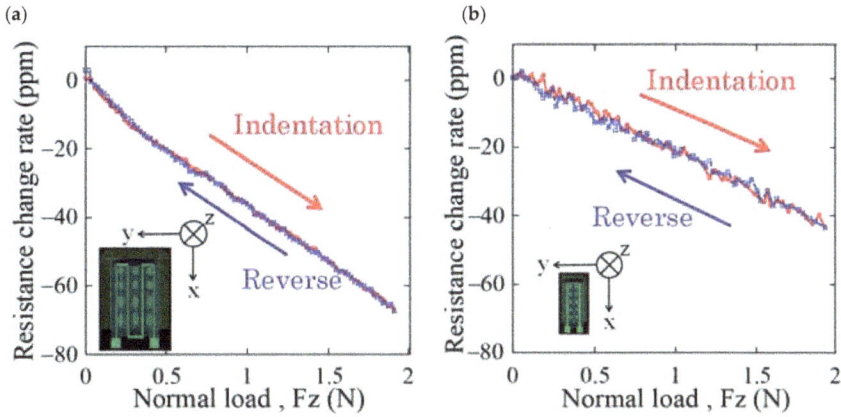

Figure 8. Response characteristic for normal load of (**a**) conventional cantilever, and (**b**) newly fabricated cantilever.

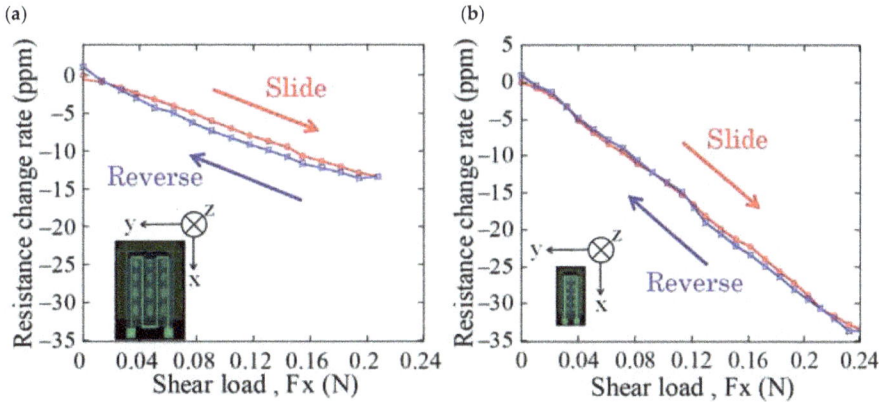

Figure 9. Response characteristic for shear load of (**a**) conventional cantilever, and (**b**) newly fabricated cantilever.

3.3. Demonstration of Tactile Sensing with High Density Cantilever Array

To demonstrate tactile sensing with cantilevers located at higher density, responses from 12 cantilevers (Type B shown in Figure 5c) to an applied shear load have been characterized. Shear load was applied uniformly on the sensor surface to direction shown in Figure 10a. Figure 10b shows responses of 12 cantilevers in the sensor Type B. The value in ppm shows resistance change rate when applied shear load is 0.24 N. Cantilevers numbered 10, 11, and 12 as shown in Figure 10a show similar response (positive resistance change) to shear load because their direction is similar to the direction of shear load. On the other hand, cantilevers 4, 5, and 6—which are located at the opposite side—show negative resistance change because their direction is opposite to the direction of shear load. Furthermore, the other cantilevers located at orthogonal direction to shear load has less response. From these results, it is found that response of the cantilever depends on relationship between directions of it and shear load. Therefore, it is demonstrated that condition of contact load distribution during gripping control can be detected using the sensor with high density cantilever array fabricated in this work.

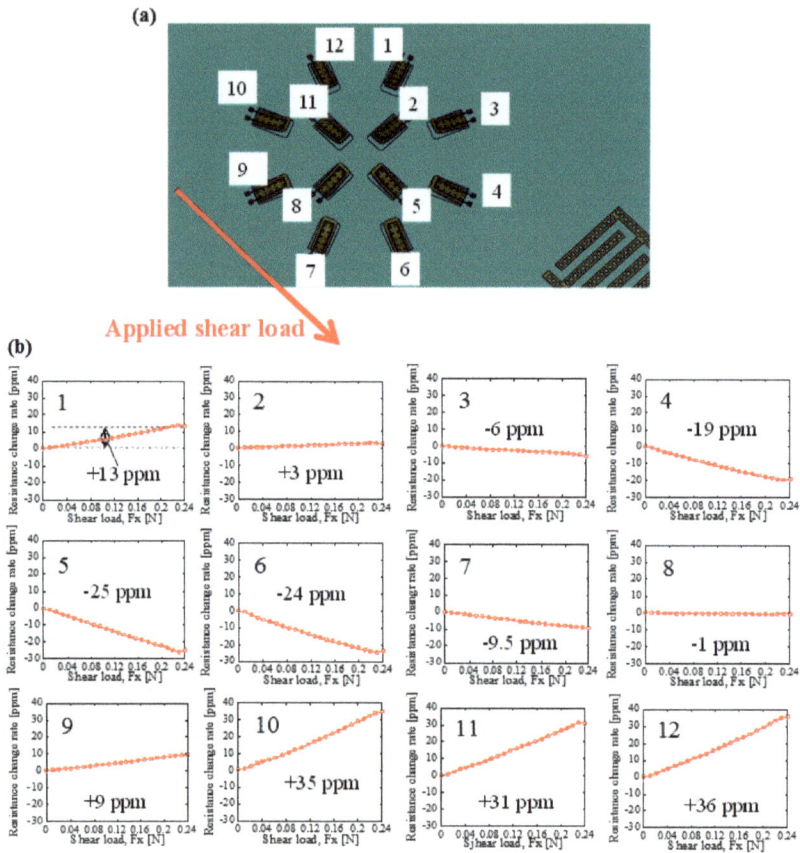

Figure 10. (a) A relationship between directions of shear load and cantilevers in Type B sensor, and (b) response of the cantilevers to shear load.

4. Conclusions

In this paper, a sensor with a smaller cantilever than the previous one and an interdigitated array electrode for optical detection was newly designed, fabricated, and evaluated. In the design of the sensor, the size of the cantilever was miniaturized to one-fifth size from previous one. Table 1 shows a comparison between the sensor fabricated in previous and this works. It is found that the optical responsivity of interdigitated array electrode with mesh holes is 1.6 times higher than that in previous works and it is demonstrated that proximity detection can be possible with high sensitivity. Furthermore, although the sensitivity to normal load of the newly fabricated cantilever is slightly smaller, the sensitivity to shear load is 2.1 times higher than previous one, and it is confirmed that the cantilever fabricated in this work has enough sensitivity to both normal and shear loads without hysteresis and detailed applied load distribution can be measured using developed sensor with high density array of 12 cantilevers. Therefore, it is expected that feedback gripping control of flexible objects is performed by detecting complicated deformation of the elastomer with higher spatial resolution. However, response sensitivities of each cantilever to applied force vector or distribution become drastically more complicated than the previous sensor with three cantilevers. In future work, a more efficient method such as application of deep-learning will be employed.

Table 1. Comparison of the structures of the previous sensor and newly fabricated sensor.

Work	Cantilever Size	Strain Gauge Resistance	Tip Height of Cantilever	Interdigitated Array Electrode	SEM Image
Previous work	Width: 200 µm, Length: 290 µm	7 kΩ	20–30 µm (±5.0 µm)	N/A	
This work	Width: 70 µm, Length: 155 µm	2 kΩ	9.60 µm (±0.7 µm)		

Author Contributions: The data of presented in this article were obtained by all authors. R.A. performed sensor designing and experiments. T.A., H.N., and M.S. contributed to preparation of experimental equipment, supervision of experiments, and design and fabrication of the sensor. In addition, all authors examined and commented on this research.

Acknowledgments: This work was supported by Grants-in-Aid for Scientific Research of Japan Society for the Promotion of Science (JP15H02230, JP17K17737), Tateisi Science and Technology Foundation, and Artificial Intelligence Research Promotion Foundation. In addition, this work was conducted as cooperative research with Koganei Corporation.

Conflicts of Interest: The authors declare no conflict of interest.

References

1. The Japan Institute for Labour Policy and Training. *Labor Situation in Japan and Its Analysis: General Overview 2015/2016*; Japan Institute for Labour Policy and Training: Tokyo, Japan, 2016; pp. 20–24.
2. Takahashi, Y.; Komeda, T.; Miyagi, M.; Koyama, H. Development of the mobile robot system to aid in the daily life for physically handicapped. In *Integration of Assistive Technology in the Information Age*; Mokhtari, M., Ed.; IOS Press: Amsterdam, The Netherlands, 2001; pp. 186–191.
3. Aravind, K.R.; Raja, P.; Pérez-Ruiz, M. Task-based agricultural mobil robots in arable farming: A review. *Span. J. Agric. Res.* **2017**, *15*, e02R01. [CrossRef]
4. Dahiya, R.; Metta, G.; Valle, M.; Sandini, G. Tactile Sensing—From Humans to Humanoids. *IEEE Trans. Robot.* **2010**, *26*, 1–20. [CrossRef]
5. Zhangping, J.; Hui, Z.; Huicong, L.; Nan, L.; Tao, C.; Zhan, Y.; Lining, S. The design and characterization of a flexible tactile sensing array for robot skin. *Sensors* **2016**, *16*, 2001.
6. Ogura, Y.; Akikawa, H.; Shimomura, K.; Morishima, A.; Lim, H.; Takanishi, A. Development of a new humanoid robot WABIAN-2. In Proceedings of the 2006 IEEE International Conference on Robotics and Automation, Orlando, FL, USA, 15–19 May 2006; pp. 76–81.
7. Noritsugu, T.; Sasaki, D.; Takaiwa, M. Application of artificial pneumatic rubber muscles to a human friendly robot. In Proceedings of the 2003 IEEE International Conference on Robotics and Automation, Taipei, Taiwan, 14–19 September 2003; Volume 3, pp. 4098–4103.
8. Hall, E.L. Intelligent robot trends and predictions for the .net future. In *Intelligent Robots and Computer Vision XX: Algorithms, Techniques, and Active Vision*; International Society for Optics and Photonics: Boston, MA, USA, 2001; Volume 4572, pp. 70–80.
9. Huebner, K.; Welke, K.; Przybylski, M.; Vahrenkamp, N.; Asfour, T.; Kragic, D.; Dillmann, R. Grasping known objects with humanoid robots: A box-based approach. In Proceedings of the 2009 International Conference on Advanced Robotics, Munich, Germany, 22–26 June 2009; pp. 1–6.
10. Johansson, R.S. 19-Sensory control of dexterous manipulation in humans. In *Hand and Brain the Neurophysiology and Psychology of Hand Movements*; Academic Press: San Diego, CA, USA, 1996; pp. 381–414.
11. Saxena, A.; Driemeyer, J.; Ng, A.Y. Robotic grasping of novel objects using vision. *Int. J. Robot. Res.* **2008**, *27*, 157–173. [CrossRef]

12. Teshigawara, S.; Tadakuma, K.; Ming, A.; Ishikawa, M.; Shimojo, M. Development of high-sensitivity slip sensor using special characteristics of pressure conductive rubber. In Proceedings of the 2009 IEEE International Conference on Robotics and Automation, Kobe, Japan, 12–17 May 2009; pp. 3289–3294.

13. Pham, A.K.; Nguyen, M.D.; Nguyen, B.K.; Phan, H.P.; Matsumoto, K.; Shimoyama, I. Multi-axis force sensor with dynamic range up to ultrasonic. In Proceedings of the 2014 IEEE 27th International Conference on Micro Electro Mechanical Systems (MEMS), San Francisco, CA, USA, 26–30 January 2014; pp. 769–772.

14. Mizoguchi, Y.; Tadakuma, K.; Hasegawa, H.; Ming, A.; Ishikawa, M.; Shimojo, M. Development of intelligent robot hand using proximity, contact and slip sensing. *Trans. Soc. Instrum. Control Eng.* **2010**, *46*, 632–640. [CrossRef]

15. Tsuji, S.; Kohama, T. A proximity and tactile sensor using self-capacitance measurement. *IEEJ Trans. Sens. Micromach.* **2014**, *134*, 400–405. [CrossRef]

16. Sohgawa, M.; Nozawa, A.; Yokoyama, H.; Kanashima, T.; Okuyama, M.; Abe, T.; Noma, H.; Azuma, T. Multimodal measurement of proximity and touch force by light- and strain-sensitive multifunctional MEMS sensor. In Proceedings of the 2014 IEEE Conference on SENSORS, Valencia, Spain, 2–5 November 2014; pp. 317–320.

17. Araki, R.; Suga, F.; Abe, T.; Noma, H.; Sohgawa, M. Gripping control of delicate and flexible object by electromotive manipulator with proximity and tactile combo mems sensor. In Proceedings of the 19th International Conference on Solid-State Sensors, Actuators and Microsystems (TRANSDUCERS), Kaohsiung, Taiwan, 18–22 June 2017; pp. 1140–1143.

18. Umeki, N.; Okuyama, M.; Noma, H.; Abe, T.; Sohgawa, M. Improvement of optical sensitivity for proximity and tactile combo sensor. *IEEJ Trans. Sens. Micromach.* **2016**, *137*, 146–150. [CrossRef]

19. Sohgawa, M.; Hirashima, D.; Moriguchi, Y.; Uematsu, T.; Mito, W.; Kanashima, T.; Okuyama, M.; Noma, H. Tactile sensor array using microcantilever with nickel–chromium alloy thin film of low temperature coefficient of resistance and its application to slippage detection. *Sens. Actuators A Phys.* **2012**, *186*, 32–37. [CrossRef]

20. Yokoyama, H.; Kanashima, T.; Okuyama, M.; Abe, T.; Noma, H.; Azuma, T.; Sohgawa, M. Active touch sensing by multi-axial force measurement using high-resolution tactile sensor with microcantilevers. *IEEJ Trans. Sens. Micromach.* **2014**, *134*, 58–63. [CrossRef]

micromachines

MDPI

Article

Reduction of Parasitic Capacitance of A PDMS Capacitive Force Sensor

Tatsuho Nagatomo [1] and Norihisa Miki [2,*

[1] School of Integrated Design Engineering, Keio University, Yokohama 223-8522, Japan;
    tatsuho19950307@keio.jp
[2] Department of Mechanical Engineering, Keio University, Yokohama 223-8522, Japan
*   Correspondence: miki@mech.keio.ac.jp; Tel.: +81-045-563-1141

Received: 27 August 2018; Accepted: 29 October 2018; Published: 3 November 2018

Abstract: Polymer-based flexible micro electro mechanical systems (MEMS) tactile sensors have been widely studied for a variety of applications, such as medical and robot fields. The small size and flexibility are of great advantage in terms of accurate measurement and safety. Polydimethylsiloxane (PDMS) is often used as the flexible structural material. However, the sensors are likely subject to large parasitic capacitance noise. The smaller dielectric constant leads to smaller influences of parasitic capacitance and a larger signal-to-noise ratio. In this study, the sensor underwent ultraviolet (UV) exposure, which changes Si–CH$_3$ bonds in PDMS to Si–O, makes PDMS nanoporous, and leads to a low dielectric constant. In addition, we achieved further reduction of the dielectric constant of PDMS by washing it with an ethanol–toluene buffer solution after UV exposure. This simple but effective method can be readily applicable to improve the signal-to-noise ratio of PDMS-based flexible capacitive sensors. In this study, we propose reduction techniques for the dielectric constant of PDMS and applications for flexible capacitive force sensors.

Keywords: polydimethylsiloxane; parasitic capacitance; ultraviolet treatment; capacitive force sensor

1. Introduction

Polymer-based flexible micro electro mechanical systems (MEMS) tactile sensors have been widely studied for a variety of applications, such as medical and robotics fields [1–11]. The small size and flexibility are of great advantage in terms of accurate measurement and safety. This trend is further accelerated by the electrodes that are robust against deformation. For example, graphene is a promising candidate and the sensors with graphene electrodes were reported along with their fabrication technologies [7,8]. Liquid metal is another candidate, which can form the electrodes by being filled up inside microchannels [9–11]. We reported flexible MEMS sensors made of polydimethyl siloxane (PDMS) structural layers and three-dimensional liquid metal electrodes to detect both normal and shear force [10,11].

PDMS is in many cases used as the structural material due to its good mechanical properties and ease of microfabrication. As is often the case with PDMS-based MEMS sensors, the parasitic capacitance originating from a short distance between wirings deteriorates the signal-to-noise ratio. In previous work, parasitic capacitance was compensated for by the electrical circuit [12,13] and low-dielectric constant materials were inserted between the wirings [14]. However, this technique drastically increases the complexity of the fabrication processes and the inserted material may deteriorate the flexibility.

In this work, to reduce parasitic capacitance, we attempted to reduce the dielectric constant of PDMS layers. PDMS is a low-k material and is known to decrease the dielectric constant after being exposed to ultraviolet (UV), which breaks up Si–CH$_3$ to increase Si–O. This topographic change makes the surface of low-k materials nanoporous, which culminates in a reduction of their dielectric constant [15,16]. In addition, we attempted to remove the unbridged materials by chemical washing in PDMS in order to make it further nanoporous. Washing PDMS with the ethanol-hexane buffer solution is known to make PDMS nanoporous [17]. We characterized the combination of these two techniques, i.e., UV-treatment and chemical washing (we used ethanol-toluene), for the first time and demonstrated reduction of the parasitic capacitance of PDMS-based sensors to enhance the signal-to-noise ratio (SNR) using the flexible sensors we developed in our prior work [11]. The techniques are simple and easily applicable to the PDMS layers, which will be stacked to form three-dimensional flexible MEMS sensors with low parasitic capacitance.

2. Theory and Method

2.1. Theory

2.1.1. Parasitic Capacitance

Parasitic capacitances are the unexpected capacitance generated between the electrical parts in a circuit. Especially in MEMS devices, the electrical parts are close together because of the small size, and therefore parasitic capacitances are difficult to ignore. Inserting low dielectric constant materials between the electrical parts was reported to be effective to reduce the parasitic capacitance [14].

2.1.2. Ultraviolet (UV) Treatment on Polydimethylsiloxane (PDMS) Porous Dielectrics

Martinez confirmed the behavior of low-k film under ultraviolet cure and measured its dielectric constant [16]. UV cure treatment breaks Si–CH$_3$ bonds and relinks Si–O bonds on the surface of low-k film, leaving nanopores. The electric constant of low-k materials has permittivity under 2.5; however, it is still higher than that of air. Therefore, nanoporous low-k films have a lower dielectric constant than nontreated low-k films. PDMS has the same Si–CH$_3$ bonds as low-k materials. PDMS is widely used as a substrate for flexible MEMS devices. We considered that UV cure treatment would lower the dielectric constant of PDMS, as shown in Figure 1, resulting in smaller parasitic capacitance.

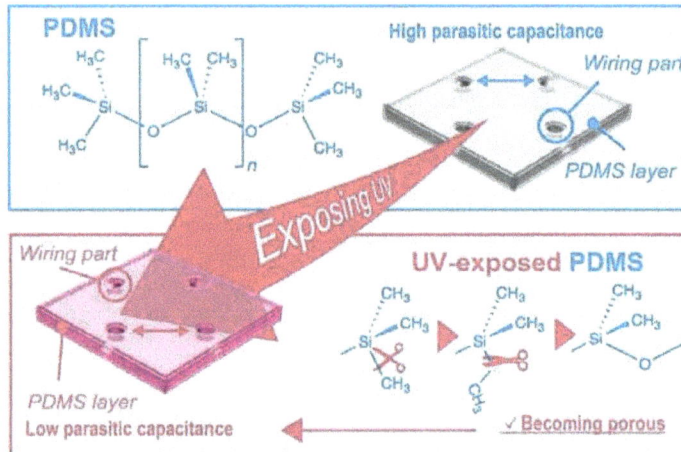

Figure 1. High parasitic capacitance caused by the wiring. The wiring will be formed inside the holes seen in the figure, which will be filled with a conductive liquid metal. Ultraviolet (UV) treatment of polydimethylsiloxane (PDMS) reduces the dielectric constant and parasitic capacitance.

2.2. Method

2.2.1. UV Treatment on PDMS Membranes

First, we fabricated PDMS membranes (Silpot 184 W/C, Dow Corning Toray Co., Ltd., Tokyo, Japan) which were spin-coated on a glass substrate with a spin coater (1H-D7, Mikasa Corporation, Hiroshima, Japan). The ratio of main agent to curing agent was 10:1. The tested film thicknesses were 250, 500, 750, and 1000 μm. Figure 2 shows the relationship between spin time and film thickness. This relationship was empirically obtained based on a power law [18–20]. After peeling off the PDMS membrane, we measured the capacitance of two parallel copper plates sandwiching each PDMS membrane by using an LCR meter (ZM2376, NF Corporation, Yokohama, Japan). The applied voltage and frequency were 1.0 V and 1.0 kHz, respectively. The dielectric constant was deduced from the measured capacitance.

Second, we exposed each PDMS membrane to UV (wavelength λ = 405 nm) for 0 to 60 s, which corresponded to 15,600 μW/m^2. After UV treatment, the dielectric constant of each membrane was deduced using the above-mentioned method.

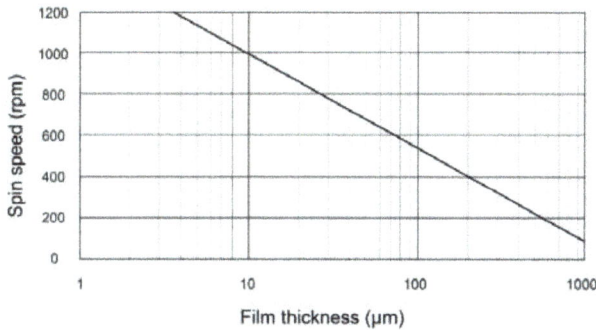

Figure 2. The relationship between spinning speed and film thickness was obtained based on a power law.

2.2.2. Washing of PDMS Membranes

Removing unbridged materials from PDMS further reduced the dielectric constant. We washed PDMS with a buffer solution. We formed a 250-μm thick PDMS membrane and exposed it to UV for 300 s, which corresponded to 78,000 μW/m^2. Subsequently, we washed the PDMS membrane for 10 s in the buffer solution, which was a mixture of ethanol (ethanol 99.5%, Fujifilm Wako Pure Chemical Corporation, Osaka, Japan) and toluene (toluene 99.5%, Fujifilm Wako Pure Chemical Corporation, Osaka, Japan) at a ratio of 10:1. The dielectric constant of the treated PDMS membrane was measured. We tested 250-μm thick membranes since it showed good results in reduction of the parasitic capacitance by UV-treatment as will be described in Section 3.1.1.

2.2.3. Periods While the Low Dielectric Constant of PDMS Membrane was Maintained

We experimentally investigated how long the low dielectric constant of PDMS membrane was maintained after treatment. We tested the UV-treated membranes and the washed UV-treated membranes. We measured the dielectric constant for 1 week with an LCR meter. The applied voltage and frequency were 1.0 V and 1.0 kHz, respectively.

2.2.4. Fabrication Process of Capacitive Force Sensor

Figure 3 shows a schematic image of the capacitive force sensor, which consisted of two parallel-plate electrodes and eight layers of liquid metal and PDMS. When a force was applied to the top of the sensor, the air pocket was deformed and the distance between the electrodes changes, thus the capacitance changes. The applied force can be deduced from the variation of capacitance.

The fabrication process is succinctly illustrated in Figure 4a. First, PDMS was poured into a poly methyl methacrylate (PMMA) mold, which was fabricated by a numerical control (NC) cutting machine (MM-100, Modia Systems Co., Saitama, Japan), and cured at 65 °C for 6 h. PDMS layers were peeled off from the molds and bonded to each other via liquid PDMS [21]; the liquid PDMS (PDMS and toluene in a 2:3 mixture) was spin-coated on a glass substrate and the bonding interface of the PDMS layers were contacted to have the liquid PDMS on the surface prior to bonding. This was cured at 65 °C for 2 h. Galinstan, a liquid metal composed of 68.5% gallium, 21.5% indium, and 10% tin, was injected into these channels. The fabricated sensor is shown in Figure 5.

Layers 6 and 7 had electrical parts that were close together. Therefore, the parasitic capacitance needed to be considered. These layers were exposed to UV light and then washed. After we fabricated PDMS layers, layers 6 and 7 were exposed to UV for 300 s (15,600–15,596 µW/m^2). We washed these layers in the buffer solution. Whole PDMS layers were bonded to each other via liquid PDMS. This was cured at 65 °C for 2 h, and finally, liquid metal was injected into these channels.

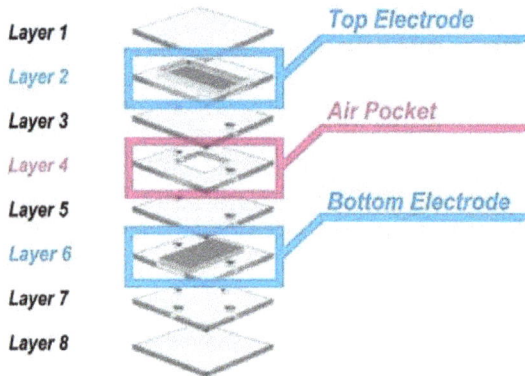

Figure 3. Structure of capacitive force sensor.

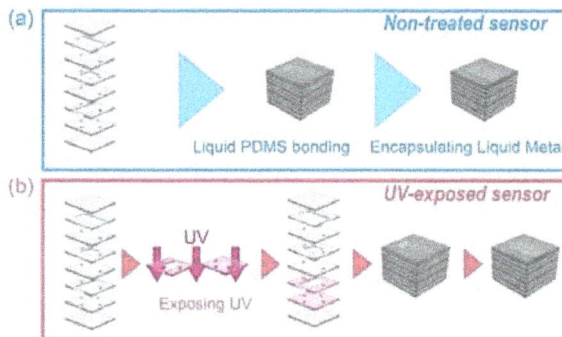

Figure 4. Fabrication process of (**a**) nontreated sensor and (**b**) UV-exposed sensor.

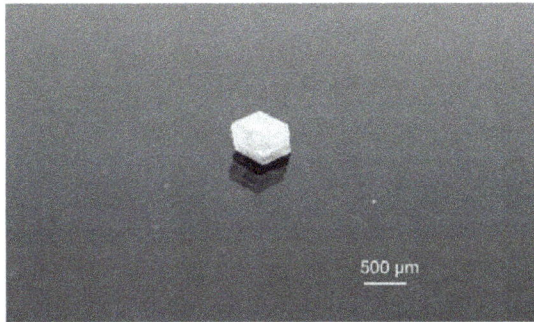

Figure 5. Photograph of the fabricated sensor.

2.2.5. Comparison with UV-Washed, UV-Treated, and Noncoated Sensors

We applied a force to the manufactured sensor of up to 1 N at a speed of 0.1 N/s using a compression machine (Micro Autograph MST-I, Shimadzu Corporation, Kyoto, Japan). The capacitance change was measured, and the influence of the parasitic capacitance was calculated using the SN ratio. In this study, we obtained the SN ratio based on 0-N–type characteristics of the Taguchi method [22]. The measurement conditions were a voltage of 1.0 V and a frequency of 1.0 kHz. This experimental setup is illustrated in Figure 6.

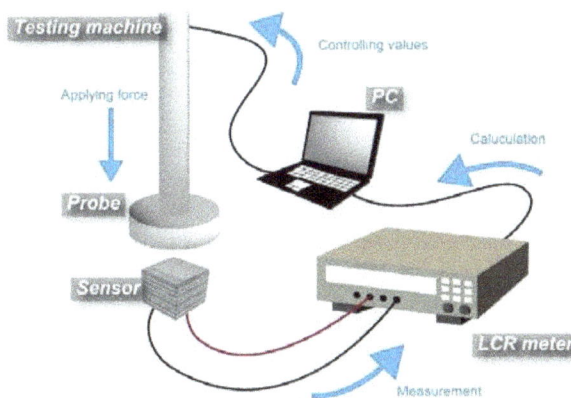

Figure 6. Experimental setup for measurement of parasitic capacitance.

3. Results and Discussion

3.1. Change of the Dielectric Constant

3.1.1. Effect of UV Treatment

Figure 7 shows the change of the dielectric constant with respect to UV exposure time. For the PDMS membrane that was 250 μm thick, a decrease of the dielectric constant was observed up to the UV exposure time of 30 s, after which the dielectric constant did not show further reduction. The PDMS membranes thicker than 500 μm showed the lowest dielectric constant after treatment for 5 s, after which the dielectric constant recovered and plateaued. These results indicate that the thin 250-μm PDMS membrane could be effectively treated using UV, while the effects became small for the thicker membranes. In low-k materials, structural changes using UV treatment were observed at the surface [15]. Therefore, the thinner PDMS membrane is more effectively modified with a larger surface/volume ratio.

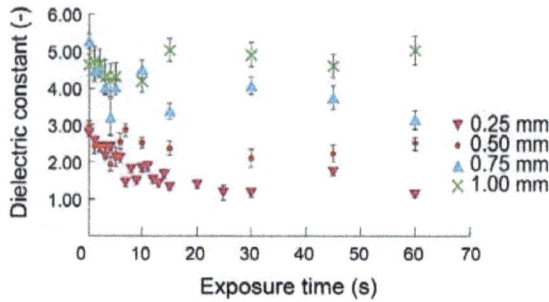

Figure 7. Dielectric constant of PDMS layers with respect to UV treatment time (n = 20).

3.1.2. Effect of Washing

We investigated the effect of washing the UV-treated PDMS membrane, as shown in Figure 8. The dielectric constant of nontreated PDMS, UV-treated PDMS, and UV-treated and washed PDMS is 2.85, 1.98, and 1.53, respectively. This result verifies the effectiveness of the washing process. It removed the unbridged materials in PDMS and promoted the formation of nanopores.

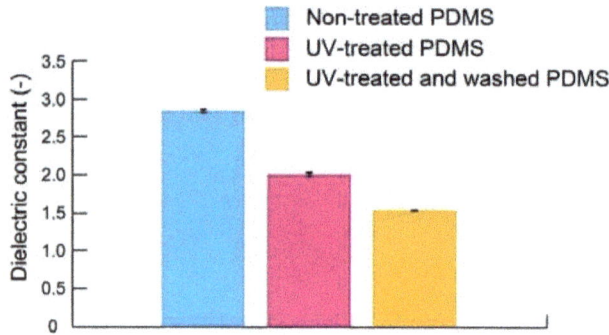

Figure 8. Dielectric constant of nontreated PDMS, UV-treated PDMS, and washed PDMS membranes (n = 20).

3.1.3. Stability of the Reduced Dielectric Constant of PDMS Membranes

We investigated the long-term stability of the effects of UV treatment and the combination of UV treatment and washing. The PDMS membrane that was 250 μm thick was exposed to UV light for 300 s. Half the samples were washed with a buffer solution for 10 s. The resulting change of the dielectric constant with respect to time is shown in Figure 9. Both samples maintained the reduced dielectric constant for 168 h, which indicates that the chemical reaction caused by UV treatment had good long-term stability. In washed PDMS, we considered that it may not return to a higher dielectric constant because the unbridged materials were removed [17]. It is assumed that Si–O groups made by UV treatment may change to Si–CH$_3$ groups included in unbridged material because of stability. However, in the case of washed PDMS, unbridged materials were already removed. Therefore, it can be said that Si–O groups cannot change to Si–CH$_3$ groups and washed PDMS may not return to a higher dielectric constant. In practical use, the sensor would be mounted onto the endoscope as a disposable part, which would require a relatively short lifetime of the sensor.

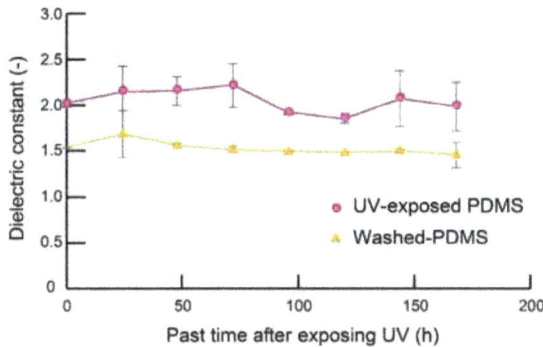

Figure 9. Dielectric constant of a 250-μm thick PDMS layer after UV treatment with or without washing (n = 20).

In this experiment, we found that the fluctuation of the dielectric constant depended on humidity. The fluctuation of the washed samples appeared to be smaller than that of the non-washed samples. We considered that this originated from the surface property of the samples. It is known that a polymer with many hydrophilic functional groups is easily affected by humidity [23–25]. After the UV treatment, the PDMS surface became hydrophilic. Washing with the buffer turned the surface back to hydrophobic, culminating in robustness against humidity and thus small fluctuations.

3.2. Sensor Characteristics

The capacitive force sensor was manufactured using PDMS membranes treated with UV for 300 s and washed with a buffer for 10 s. The manufacturing process took 6 h, therefore the low dielectric constant was considered to be maintained. Figure 10 shows the SN ratio of the sensor using PDMS membranes that underwent no treatment, UV-treatment, and a combination of UV treatment and washing. The UV treatment increased the SN ratio from 5.4 to 5.6 in dB, the washing treatments increased the SN ratio from 5.6 to 6.2 in dB, i.e., approximately 1.2 times. The wiring part had parasitic capacitance because parasitic capacitance is caused by close wiring. Therefore, we replaced layers 6 and 7 with low-dielectric PDMS layers in order to reduce parasitic capacitance. It can be said that we can successfully reduce parasitic capacitance by stacking multiple thin low-dielectric PDMS layers. This result indicates that treatment of PDMS with UV and washing reduces the dielectric constant of the PDMS membrane structures and thus improves the performance of the capacitive sensor.

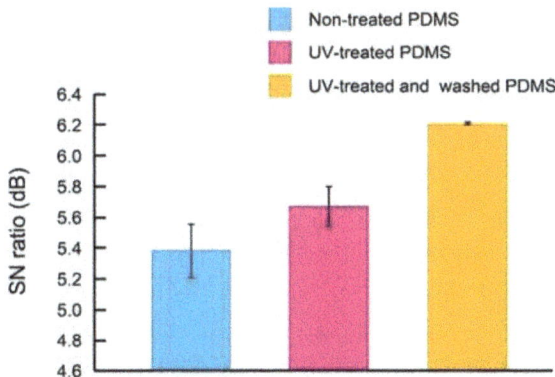

Figure 10. SN ratio obtained from nine devices, the nontreated devices, UV-treated devices, and the UV-exposed and washed devices (n = 3).

4. Conclusions

We experimentally confirmed that the furthest reduction of dielectric constant of PDMS achieved by the combination of treatment and washing with the ethanol-toluene buffer solution for the first time. We applied the proposed combination to the 3-D flexible sensor that we reported in our prior work and successfully enhanced the SN ratio from 5.4 to 6.2 dB. The proposed technique is simple but effective and can be readily applicable to PDMS-based flexible sensors.

Author Contributions: Conceptualization, T.N.; methodology, T.N.; validation, T.N. and N.M.; formal analysis, T.N.; investigation, T.N. and N.M.; resources, N.M.; data curation, T.N.; writing—original draft preparation, T.N. and N.M.; writing—review and editing, N.M.; visualization, T.N.; supervision, N.M.; project administration, N.M; funding acquisition, N.M.

Funding: This work was funded by JSPS KAKENHI, grant number 18H03276.

Conflicts of Interest: The authors declare no conflict of interest.

References

1. Tiwana, M.I.; Redmond, S.J.; Lovell, N.H. A review of tactile sensing technologies with applications in biomedical engineering. *Sens. Actuators Phys.* **2012**, *179*, 17–31. [CrossRef]
2. Hu, Y.; Katragadda, R.B.; Tu, H.; Zheng, Q.; Li, Y.; Xu, Y. Bioinspired 3-D Tactile Sensor for Minimally Invasive Surgery. *J. Microelectromechanical Syst.* **2010**, *19*, 1400–1408. [CrossRef]
3. Yin, J.; Santos, V.J.; Posner, J.D. Bioinspired flexible microfluidic shear force sensor skin. *Sens. Actuators Phys.* **2017**, *264*, 289–297. [CrossRef]
4. McKinley, S.; Garg, A.; Sen, S.; Kapadia, R.; Murali, A.; Nichols, K.; Lim, S.; Patil, S.; Abbeel, P.; Okamura, A.M.; et al. A single-use haptic palpation probe for locating subcutaneous blood vessels in robot-assisted minimally invasive surgery. In Proceeding of the 2015 IEEE International Conference on Automation Science and Engineering (CASE), Gothenburg, Sweden, 24–28 August 2015.
5. Engel, J.; Chen, J.; Liu, C. Development of polyimide flexible tactile sensor skin. *J. Micromech. Microeng.* **2003**, *13*, 359–366. [CrossRef]
6. Nishio, N.; Takanashi, H.; Tetsushi, M. Development of Hardness Sensor using Strain Gauge. In Proceeding of the Society of Instrument and Control Engineers Tohoku Chapter (SICE), Akita, Japan, 9 July 2010; Volume 259, pp. 1–5.
7. Chen, Z.; Ren, W.; Gao, L.; Liu, B.; Pei, S.; Cheng, H.-M. Three-dimensional flexible and conductive interconnected graphene networks grown by chemical vapour deposition. *Nat. Mater.* **2011**, *10*, 424–428. [CrossRef] [PubMed]
8. Samad, Y.A.; Li, Y.; Schiffer, A.; Alhassan, S.M.; Liao, K. Graphene Foam Developed with a Novel Two-Step Technique for Low and High Strains and Pressure-Sensing Applications. *Small* **2015**, *11*, 2380–2385. [CrossRef] [PubMed]
9. Ota, H.; Chen, K.; Lin, Y.; Kiriya, D.; Shiraki, H.; Yu, Z.; Ha, T.-J.; Javey, A. Highly deformable liquid-state heterojunction sensors. *Nat. Commun.* **2014**, *5*. [CrossRef] [PubMed]
10. Nakadegawa, T.; Ishizuka, H.; Miki, N. Three-axis scanning force sensor with liquid metal electrodes. *Sens. Actuators Phys.* **2017**, *264*, 260–267. [CrossRef]
11. Nagatomo, T.; Miki, N. Three-axis capacitive force sensor with liquid metal electrodes for endoscopic palpation. *IET Micro Nano Lett.* **2017**, *12*, 564–568. [CrossRef]
12. Zhang, P.; Wan, Q.; Feng, C.; Wang, H. All Regimes Parasitic Capacitances Extraction Using a Multi-Channel CBCM Technique. *IEEE Trans. Semicond. Manuf.* **2017**, *30*, 121–125. [CrossRef]
13. Sun, L.-J.; Cheng, J.; Ren, Z.; Shang, G.-B.; Hu, S.-J.; Chen, S.-M.; Zhao, Y.-H.; Zhang, L.; Li, X.-J.; Shi, Y.-L. Extraction of geometry-related interconnect variation based on parasitic capacitance data. *IEEE Electron Device Lett.* **2014**, *35*, 980–982. [CrossRef]
14. Hebedean, C.; Munteanu, C.; Racasan, A.; Pacurar, C. Parasitic capacitance removal with an embedded ground layer. In Proceedings of the Eurocon 2013, Zagreb, Croatia, 1–4 July 2013; IEEE: Zagreb, Croatia, 2013; pp. 1886–1891.
15. Martinez, E.; Rochat, N.; Guedj, C.; Licitra, C.; Imbert, G.; Le Friec, Y. Influence of electron-beam and ultraviolet treatments on low-k porous dielectrics. *J. Appl. Phys.* **2006**, *100*, 124106. [CrossRef]

16. Kao, K.-C.; Chang, W.-Y.; Chang, Y.-M.; Leu, J.; Cheng, Y.-L. Effect of UV curing time on physical and electrical properties and reliability of low dielectric constant materials. *J. Vac. Sci. Technol. Vac. Surf. Films* **2014**, *32*, 061514. [CrossRef]

17. Miwa, S.; Ohtake, Y. Chemical Changes in Cross-linked Silicone Rubber by Ozone-water Treatments. *Nippon Gomu Kyokaishi* **2014**, *87*, 161–167. [CrossRef]

18. Meyerhofer, D. Characteristics of resist films produced by spinning. *J. Appl. Phys.* **1978**, *49*, 3993–3997. [CrossRef]

19. Haas, D.E.; Quijada, J.N.; Picone, S.J.; Birnie, D.P. Effect of solvent evaporation rate on skin formation during spin coating of complex solutions. *Proc. SPIE* **2000**, *3943*, 280–284.

20. Emslie, A.G.; Bonner, F.T.; Peck, L.G. Flow of a Viscous Liquid on a Rotating Disk. *J. Appl. Phys.* **1958**, *29*, 858–862. [CrossRef]

21. Wu, H.; Huang, B.; Zare, R.N. Construction of microfluidic chips using polydimethylsiloxane for adhesive bonding. *Lab Chip* **2005**, *5*, 1393. [CrossRef] [PubMed]

22. Taguchi, G. Quality Engineering (Taguchi Methods) For The Development Of Electronic Circuit Technology. *IEEE Trans. Reliab.* **1995**, *44*, 225–229. [CrossRef]

23. Anderson, J.H.; Parks, G.A. Electrical conductivity of silica gel in the presence of adsorbed water. *J. Phys. Chem.* **1968**, *72*, 3662–3668. [CrossRef]

24. Bodas, D.; Khan-Malek, C. Formation of more stable hydrophilic surfaces of PDMS by plasma and chemical treatments. *Microelectron. Eng.* **2006**, *83*, 1277–1279. [CrossRef]

25. Mata, A.; Fleischman, A.J.; Roy, S. Characterization of Polydimethylsiloxane (PDMS) Properties for Biomedical Micro/Nanosystems. *Biomed. Microdevices* **2005**, *7*, 281–293. [CrossRef] [PubMed]

micromachines

MDPI

Article

Fatigue Assessment by Blink Detected with Attachable Optical Sensors of Dye-Sensitized Photovoltaic Cells

Ryogo Horiuchi, Tomohito Ogasawara and Norihisa Miki *

Department of Mechanical Engineering, Keio University, 3-14-1 Hiyoshi, Kohoku-ku, Yokohama, Kanagawa 223-8522, Japan; jjryogo747@gmail.com (R.H.); tomo.0602.72003@gmail.com (T.O.)
* Correspondence: miki@mech.keio.ac.jp; Tel.: +81-45-566-1430

Received: 11 May 2018; Accepted: 14 June 2018; Published: 20 June 2018

Abstract: This paper demonstrates fatigue assessment based on eye blinks that are detected by dye-sensitized photovoltaic cells. In particular, the sensors were attached to the temple of eyeglasses and positioned at the lateral side of the eye. They are wearable, did not majorly disturb the user's eyesight, and detected the position of the eyelid or the eye state. The optimal location of the sensor was experimentally investigated by evaluating the detection accuracy of blinks. We conducted fatigue assessment experiments using the developed wearable system, or smart glasses. Several parameters, including the frequency, duration, and velocity of eye blinks, were extracted as fatigue indices. Successful fatigue assessment by the proposed system will be of great benefit for maximizing performance and maintenance of physical/mental health.

Keywords: fatigue; dye-sensitized photovoltaic cells; wearable; blink; sensors; micro/nano technology; microfabrication

---

1. Introduction

Fatigue assessment is crucial to secure safety and efficiency in operation. For such applications, the assessment system itself should provide the users' minimum physical and mental stress; the whole system should be light enough to be wearable, and should not disturb the users' activities and eyesight. Real-time process is also an important requirement, which encourages us to discover fatigue indices that can be easily measured as well as processed.

Heart rate variance (HRV) uses an R-R interval, or RRI, of an electrocardiogram (ECG). The RRI is deduced from the measured ECG and Fourier transformed to calculate the autonomic nerve index as the power ratio of the low-frequency (0.05–0.15 Hz) and high-frequency (0.15–0.40 Hz) bands. Heart rate variance increases with fatigue and decreases during recovery [1,2].

Electroencephalograms (EEGs), or brain waves, are reported to represent fatigue [3–6]. The relationship between EEG and mental fatigue has been explored by many reports, where they attempted to find effective indexes of fatigue from EEGs. Four-frequency bands were often used, including alpha (8–13 Hz), beta (17–34 Hz), theta (4–8 Hz), and delta (0.5–4 Hz). The peak frequency of the alpha power decreases with loaded mental work and increases with resting [3]. The ratio of the low alpha (8–10.5 Hz) to the high alpha (10.5–13 Hz) was experimentally verified to better represent the mental work load [4]. The other bands, beta, theta, and a combination of all the bands, were correlated to fatigue [5–10].

Information acquired from eyes, including movement of the eye and eye blinks, provides the state of the subjects [11–13]. In order to obtain the information with minimal physical and mental stress, we used a wearable, see-through eyeglass-type eye-tracking system [14,15]. The system had an array of transparent optical sensors, which were dye-sensitized photovoltaic devices [16–18].

While dye-sensitized photovoltaic devices have been studied as next generation solar cells, we utilized them as optical sensors. In this context, the sensor property we were most interested in was the detection accuracy of eye movement and eye blinks. The properties of the dye-sensitized photovoltaic devices, such as photo conversion efficiency, wavelength dependence, etc., are not discussed. The sensor system was wearable and light, and did not have cameras pointing at the user, and therefore, provided little physical and mental stress. In our prior work, the indices related to eye blinks were found to reflect the fatigue of the subjects [15]. In this work, in order to further reduce stress to the users, we designed and fabricated a system to detect eye blinks, which can be attached to the temple of the eyeglasses and positioned at the lateral side of the face, as shown in Figure 1. It does not hinder the eyesight of the users, and is not affected by the movement of the eye. It can be attached to various types of eyeglasses and eyeglass-type devices. We firstly deduced the optimal position of the blink detection system experimentally and then attempted to correlate the users' fatigue with the eye blinks that were detected by the system.

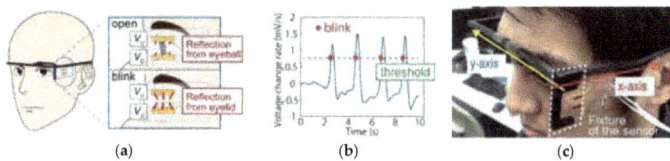

(a) (b) (c)

Figure 1. Smart glass system to detect blinks. (**a**) Schematic view of the sensor cells. Two dye-sensitized photovoltaic cells were patterned on a glass substrate as the optical sensors, which detect the reflection light from the eye. The reflection light from the eyelid when the subject blinks is larger than that from the eyeball when he/she opens the eye. (**b**) Detection of the blink. When the derivative of the average output of the two sensor cells exceeds the threshold, we consider the subject to be blinking. (**c**) Position of the system. The positions in the *x*- and *y*-axes can be varied by the location to set the fixture onto the temple and by the holder design, respectively.

2. Materials and Methods

2.1. Blink Detection System

The blink detection system is composed of two dye-sensitized photoelectric cells, as shown in Figure 1. The cell is patterned on a glass substrate 2 mm in width and 8 mm in length. The detailed fabrication processes were described in prior work [17] and in Figure S1 in the Supplemental Material. Indium-Tin~Oxide (ITO) thin film on the glass substrate was fine-patterned to form an electrical circuit. A titanium dioxide nanoparticles film was patterned to form the cathode, on which ruthenium dyes adsorb. Fabrication processes of the cells were completed with encapsulation of the electrolyte between the cathode and the anode on the glass substrates. The fabricated cell was attached to the temple of the eyeglasses. The sensors detect the reflection light from the eyelid and the eyeball. The reflection light from the eyeball is weaker than that from the eyelid. Thus, the sensors can detect the eye-state including blinks.

In the blink detection, the derivative of the average of the output voltages of the two cells (V_U and V_D) is used, as shown in Figure 1a. Blink is detected when the derivative with respect to time, or voltage change rate, is greater than the threshold. In our prior work, we investigated the successful detection of the eye blinks with respect to the threshold. Based on the experimental results, we set the threshold to be 70% of the maximum voltage change rate [14,15], as illustrated in Figure 1b. The optimal position of the cells to have the most reliable detection was experimentally deduced. The position of the system in *x*- and *y*-axes were varied, with the position of the fixture at the temple and the fixture design, as shown in Figure 1c. Four subjects (21~24 years old, 3 males and 1 female) were requested to wear the smart glass system and blink every 2 s. The number of blinks detected by the system was compared to that obtained by an external camera. The average of the voltage change rate at the blinks was defined as the detection sensitivity (mV/s) and used to find the optimal position of the system. Each subject conducted the experiments four times.

2.2. Fatigue Assessment Experiments

The experiments described in this work were approved by the research ethics committee of the Faculty of Science and Technology, Keio University. Sixteen subjects (aged 21~24, 13 males and 3 females) participated in the experiments.

The subjects were directed to perform the Uchida–Kraepelin (U–K) psychodiagnostic test, which was originally developed by Uchida [19]. The U–K test is based on a series of simple addition tasks and measures a subject's ability to perform tasks quickly and accurately. It is also used as a simple but efficient mental stressor.

We attempted to assess the fatigue by three methods: subjective fatigue symptoms or "jikaku-sho shirabe" in Japanese, HRV, and the blinks. Subjective fatigue symptoms were deduced by a questionnaire for 25 symptoms [20–22]. The subjects were requested to score 1–5 for each symptom, where 1 is totally disagree and 5 is totally agree. This protocol was proposed by the Industrial Fatigue Research Committee of Japanese Occupational Health in 2002. The symptoms are categorized into five factors, which are (i) drowsiness, (ii) instability, (iii) uneasiness, (iv) local pain or dullness, and (v) eyestrain. These factors were correlated with the number of U–K tests.

Heart rate variability was measured with flat electrodes from the arm and leg of the subject and then acquired by a polygraph. The signals were processed by Labchart (ADInstruments, Nagoya, Japan), where power supply noise at 50 Hz was cut and signals from 0.5 to 10 Hz were passed. In the power spectrum analysis of HRV, the ratio of the low-frequency power to the high-frequency power was deduced to objectively and quantitatively assess the fatigue induced by the U–K tests.

Blinks by the subjects were measured using the smart glass system. We investigated (i) the number of blinks, (ii) the number of blink bursts, (iii) the blink burst rate, (iv) the blink duration, and (v) blink velocity as the candidates for the fatigue indices. The blink bursts were series of blinks between 0.5~2.0 s. The blink duration represents the period while the eyes are closed. The blink velocity is the velocity of the eyelids. All the candidates can be deduced from the output signals of the smart glass system.

In analyses of subjective fatigue symptoms, HRV, and the blinks, the obtained data were normalized as follows.

$$z_i = \frac{x_i - \widetilde{x}}{\sigma^2} \tag{1}$$

where x_i is the acquired datum, $\widetilde{x}$ is the average, and σ^2 is the variance. The data group has an average of 0 and a standard deviation of 1.

The protocol of the fatigue assessment experiments is illustrated in Figure 2. The fatigue symptom questionnaire and HRV measurement were conducted for 3 min, which was followed by U–K tests and blink detection for 8 min. This set was iterated five times, and the series of experiments ended with another fatigue symptom questionnaire and HRV measurement.

The obtained data are not likely to be normally distributed. We conducted Steel–Dwass multiple comparison tests to investigate the indices with respect to the number of U–K tests or fatigue.

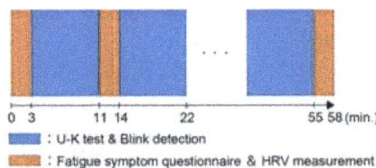

0 3   11 14   22   55 58 (min.)
▮ : U-K test & Blink detection
▮ : Fatigue symptom questionnaire & HRV measurement

Figure 2. Protocol of the fatigue assessment experiments.

3. Results

3.1. Optimal Position of the Blink Detection System

Figure 3a shows the photo of the fabricated smart glass system. Two dye-sensitized photovoltaic cells are formed on a glass substrate as the optical sensors, which is attached to the temple of the eyeglasses, which is suitable for wearable applications. The position of the glass substrate can be varied in x- and y-axes as shown in Figures 1c and 3b. The position lateral to the pupil is set to be $x = 0$.

The detection sensitivity, which is the derivative of the voltage at the blinks, with respect to x-axis position and y-axis position is shown for four subjects in Figure 4a,b, respectively. At the x-axis, for all the subjects, the sensitivity showed its maximum at $x = 0$. This is reasonable since the motion of the eyelid is the largest at $x = 0$. We were concerned about the effect of eyelashes, however, it did not appear in the experiments. Differences among the subjects in absolute values of the sensitivity were observed, which were calibrated in the following blink detection experiments.

At the y-axis, the sensitivity increased as the sensor was closer to the eye. Since the sensor detects the scattered light from the eyelid, the sensor output becomes larger as the gap between the sensor and eyelid decreases. In the following experiments, we set the smart glass system such that the distance from the eye was as small as possible while the sensor did not touch the lateral side of the head.

The blink detection accuracy was deduced by the ratio of the blinks detected by the smart glass system to those detected by an external video camera image. The experiments were conducted for four subjects, and the average and standard deviation is shown in Figure 5. It was found to be as high as 94% and showed no significant difference (p-value > 0.05) from our previous device [15].

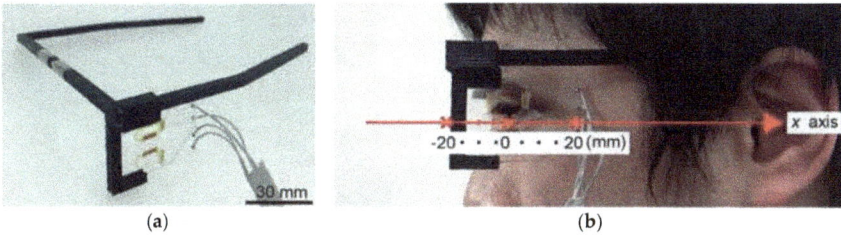

Figure 3. (a) Photo of the fabricated smart glass system. (b) Photo of the experiments highlighting the position of the system at the x-axis.

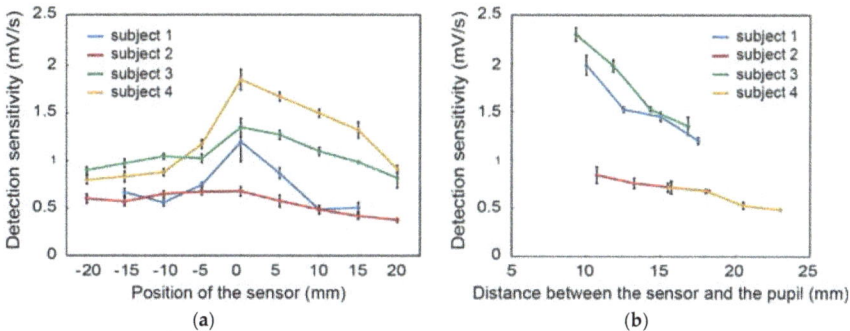

Figure 4. Detection sensitivities with respect to the sensor position at the (a) x-axis and (b) y-axis. Four subjects conducted experiments four times for each condition. The optimal position on the x-axis was found to be $x = 0$ (i.e., the lateral to the eye). On the y-axis, the sensitivity increased as the sensor was located closer to the eye. The error bars represent the standard deviation.

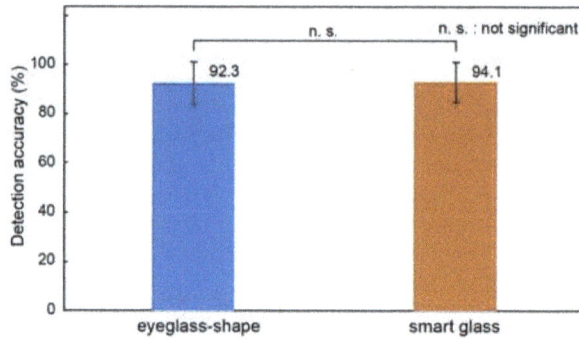

Figure 5. Blink detection accuracy. The results obtained by the smart glass system were compared to the eyeglass system in our prior work and showed no significant difference. Four subjects participated in the experiments for each device.

3.2. Fatigue Assessment Experiments

3.2.1. Subjective Fatigue Symptoms

Figure 6 shows the normalized results of (a) drowsiness, (b) instability, (c) uneasiness, (d) local pain or dullness, and (e) eyestrain, with respect to the number of U–K tests. For all the symptoms, significant differences between before (experiment number 0) and after all the U–K tests (number 5) were obtained. The figures also show significant differences in the results at each experiment from before the U–K tests, if any. The results indicate that the applied U–K tests induced fatigue to the subjects, which increased with the number of tests.

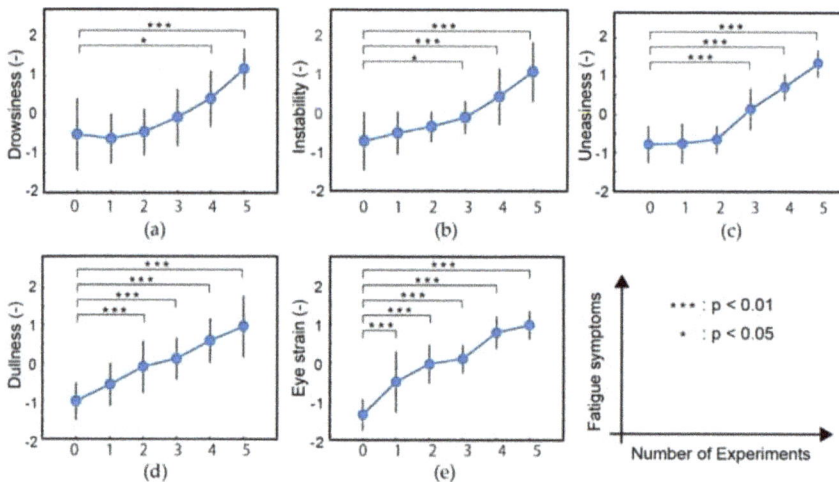

Figure 6. Normalized (**a**) drowsiness, (**b**) instability, (**c**) uneasiness, (**d**) local pain or dullness, and (**e**) eyestrain with respect to the number of U–K tests ($n = 16$). "p" represents the p-value in statistics hypothesis testing.

Of the five symptoms, eye strain appears to correlate with the number of experiments most. Eye strain increases with the number of tests from the first to the final test. Drowsiness, instability, and uneasiness increased from the third test. Since fatigue is the summary of these symptoms, we cannot say that fatigue increases with the number of tests almost linearly, as shown in the eye strain, or fatigue does not appear in the first two tests and starts increasing from the third test. However, at least we can conclude that fatigue increased as the U–K tests were iterated, and this trend was in common among all 16 subjects with relatively small deviation. It is reasonable to presume fatigue accumulates from the beginning of the tests, although it may be little. If we would like to detect such small amounts of fatigue, eye strain is the most suitable symptom to investigate, which can be represented by the blinks.

3.2.2. HRV

Figure 7 shows the normalized HRV with respect to the number of U–K tests. Large variations among the subjects in HRV before the U–K tests was found. As an overall trend, HRV increased with the number of tests. Significant differences were found between after the first test and the final test. However, the trend appears to be less definitive than the subjective evaluation, as shown in Figure 6. In our other work [5], HRV did not match well with the subjective evaluation, either. Heart rate variance may provide qualitative information on fatigue, but it is not suitable to quantitatively assess fatigue.

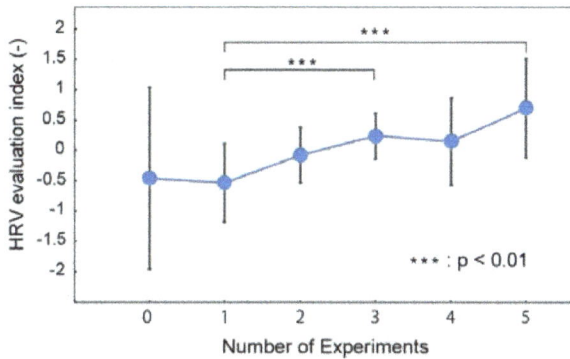

Figure 7. Normalized heart rate variance (HRV) with respect to the number of U–K tests ($n = 16$). "*p*" represents the *p*-value in statistics hypothesis testing.

3.2.3. Blinks

We investigated (i) the number of blinks, (ii) the number of blink bursts, (iii) the blink burst rate, (iv) the blink duration, and (v) blink velocity as candidates for the fatigue indices. The results are shown in Figure 8.

The number of blinks increased monotonously with the number of tests, as shown in Figure 8a. Significant difference ($p < 0.01$) was found even between the 1st and 2nd tests. The trend is similar with the eye strain (Figure 6e). Given this agreement and rather easy measurement, the number of blinks is a strong candidate for the fatigue index.

As shown in Figure 8b,c, blink bursts showed good correlation with the number of blinks. A significant difference was found from the 2nd tests, which represents the high sensitivity with fatigue similarly with the number of blinks.

Blink duration, on the contrary, did not show good correlation with the number of the U–K tests or fatigue, as shown in Figure 8d. The duration was reported as a fatigue index in prior work [23]. It was also reported that the duration did not extend when the level of awakeness was too low. We considered that some subjects might have a low level of awakeness during the tests. The level of awakeness can be evaluated by the activity of the parasympathica divisionis, which is reflected in the high frequency component of the heart rate. We investigated the blink duration of the subjects whose high frequency component of the heart rate did not increase. The results are shown in Figure 9. The blink duration increased as a trend. However, we consider that the blink duration cannot be a good fatigue index because of the low correlation with the fatigue and limitation of the subjects.

Figure 8e shows the normalized blink velocity with respect to the number of tests. The velocity decreased with the tests. A significant difference was found between the 1st and 3rd test. However, the variation among the subjects was found to be larger and the correlation with the number of tests was smaller than the number of blinks and blink bursts.

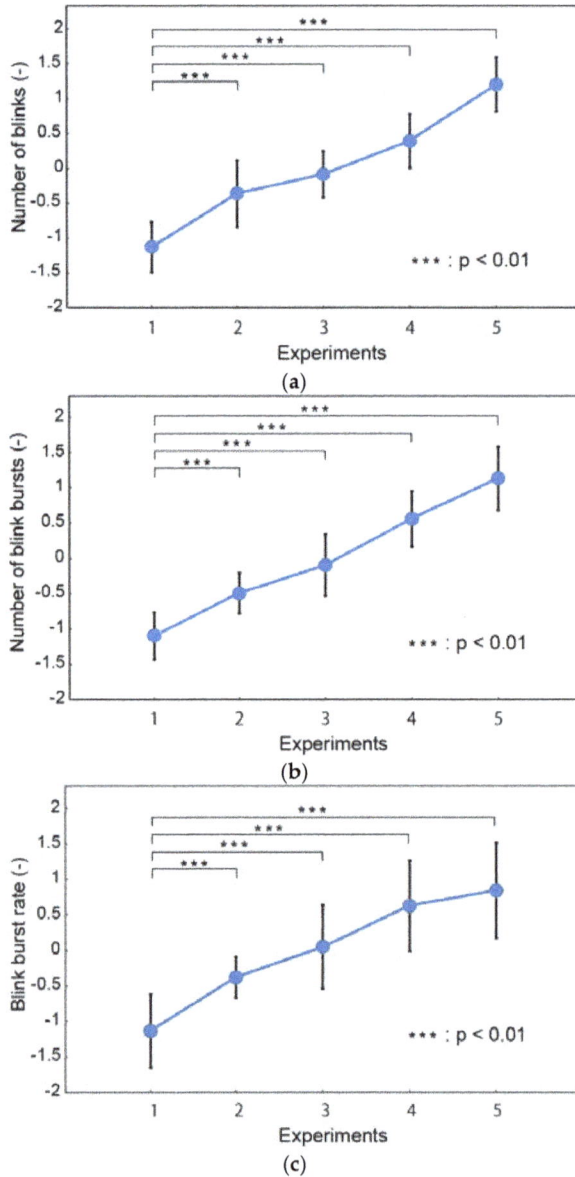

(a)

(b)

(c)

Figure 8. *Cont.*

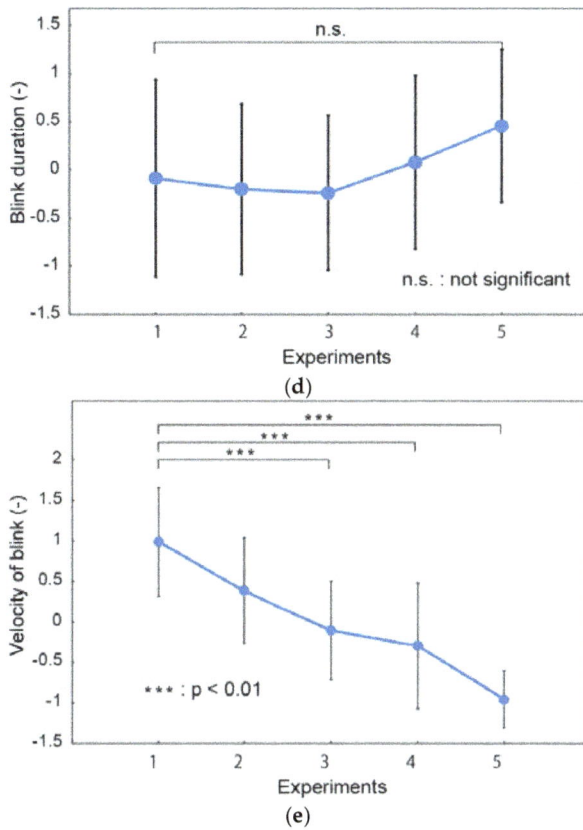

Figure 8. Obtained (**a**) the number of blinks, (**b**) the number of blink bursts, (**c**) the blink burst rate, (**d**) the blink duration, and (**e**) blink velocity with respect to the number of U–K tests. "*p*" represents the *p*-value in statistics hypothesis testing.

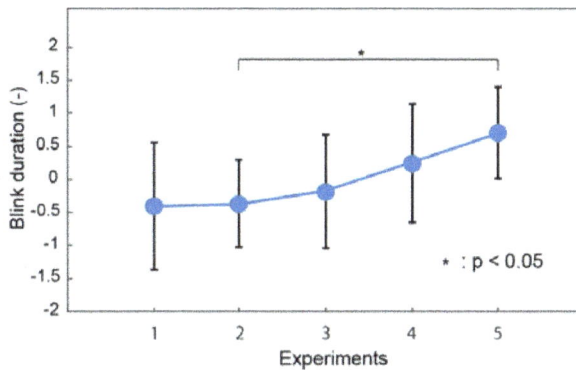

Figure 9. Blink duration for the subjects who did not have a low level of awakeness. "*p*" represents the *p*-value in statistics hypothesis testing.

4. Discussion

The number of blinks was found to be the most suitable index to assess fatigue. However, note that it is reasonable to presume the number of blinks will hit its maximum value while fatigue can keep increasing. In our experiments, the tests were conducted for approximately one hour. For this fatigue level, the number of blinks is a good index. In our future work, we will investigate the range of fatigue level, where the number of blinks maintains good correlation with fatigue.

Table 1 summarizes the results obtained in our prior work with an eyeglass-type system [15] and this work with the smart glass. The indices obtained by the smart glass system were found to be better than those obtained by the eyeglass system. As we showed in Figure 5, the blink detection accuracy was found to be comparable. However, during the U–K tests, the subjects move their eyes, which affects the detection accuracy more dominantly for the eyeglass type than the smart glasses. Figure 10 shows the sensor output data when the subject moves his/her eye with an angle of $10°$ and blinks with the eye movement. Obviously, detection of blinks with the eyeglasses-type device is disturbed by the eye movement. We conclude that the smart glass system proposed herein is more suitable to assess fatigue than our previous device.

Table 1. Summary of the candidates for fatigue index.

Blink Detection Sensor	Number of Blinks	Number of Blink Bursts	Blink Bursts Rate	Blink Duration	Velocity of Blinks
Eyeglasses Type [15]	†	†	†	n.s.	–
Smart Glass Type	****	****	****	†	****

Notes: ****: $p < 0.001$; †: $0.05 < p < 0.1$; n.s.: not significant.

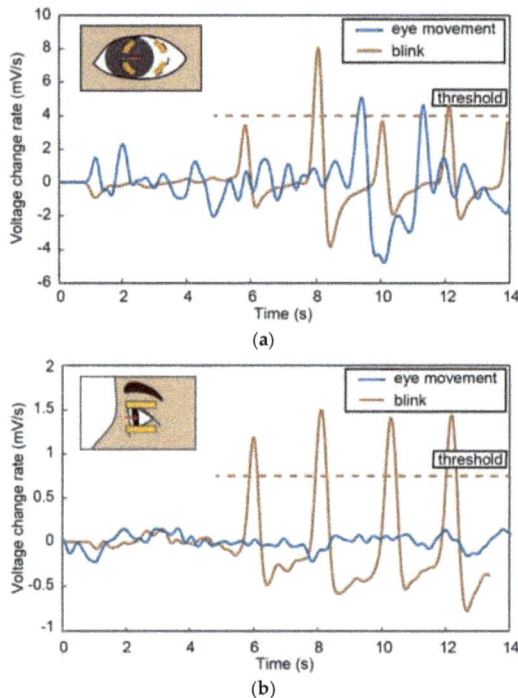

Figure 10. The sensor outputs when the subject moves the eye and blinks while moving the eye in case of (**a**) eyeglasses type device and (**b**) smart glass device.

Micromachines **2018**, *9*, 310

5. Conclusions

Fatigue assessment using the proposed sensor system was successfully demonstrated. The number of blinks and blink bursts showed good correlation with the number of U–K tests, which also showed good agreement with the fatigue symptoms that were deduced from subjective questionnaires. The measurement of the number of blinks can be achieved via simple processes, which can contribute to real-time fatigue assessment. Heart rate variability, or HRV, showed less correlation with the tests. It was experimentally found that the sensor system attached to the temple of the eyeglasses, or smart glass system, was more suitable than the eyeglasses-type system due to the little disturbance from eye movement. Fatigue assessment by the proposed smart glass system is of great benefit for maximizing performance and maintenance of physical/mental health.

Supplementary Materials: The following are available online at http://www.mdpi.com/2072-666X/9/6/310/s1.

Author Contributions: Conceptualization, R.H. and N.M.; Methodology, R.H. and N.M.; Validation, R.H., T.O., and N.M.; Formal Analysis, R.H.; Investigation, R.H. and T.O.; Resources, N.M.; Data Curation, R.H.; Writing-Original Draft Preparation, R.H. and N.M.; Writing-Review & Editing, N.M.; Visualization, R.H.; Supervision, N.M.; Project Administration, N.M.; Funding Acquisition, N.M.

Funding: This research was funded by JSPS KAKENHI Grant Number 18H03276.

Conflicts of Interest: The authors declare no conflict of interest.

References

1. Sloan, R.P.; Shapiro, P.A.; Bagiella, E.; Boni, S.M.; Paik, M.; Bigger, J.T., Jr.; Steinman, R.C.; Gorman, J.M. Effect of mental stress throught the day on cardiac autonomic control. *Biol. Psychol.* **1994**, *37*, 89–99. [CrossRef]
2. Kim, H.G.; Cheon, E.J.; Bai, D.S.; Lee, Y.H.; Koo, B.H. Stress and Heart Rate Variability: A meta-analysis and review of the literature. *Psychiatry Investig.* **2018**, *15*, 235–245. [CrossRef] [PubMed]
3. Tanaka, M.; Ishii, A.; Watanabe, Y. Effects of mental fatigue on brain activity and cognitive performance: A magnetoencephalography study. *Anat. Physiol.* **2015**, *s4*, 1–5. [CrossRef]
4. Shimizu, N.; Saito, T.; Hukumoto, K. Research of the mental fatigue reduction effect color-light environment control. *Res. Rep. Nagaoka Univ. Technol.* **2003**, *25*, 87–91.
5. Kudo, Y.; Arai, M.; Miki, N. Fatigue assessment by electroencephalogram measured with candle-like dry microneedle electrodes. *Micro Nano Lett.* **2017**, *12*, 545–549. [CrossRef]
6. Ray, W.J.; Cole, H.W. EEG alpha activity reflects attentional demands, and beta activity reflects emotional and cognitive processes. *Science* **1985**, *228*, 750–752. [CrossRef] [PubMed]
7. Craig, A.; Tran, Y.; Wijesuriya, N.; Nguyen, H. Regional brain wave activity changes associated with fatigue. *Psychophysiology* **2012**, *49*, 574–582. [CrossRef] [PubMed]
8. Grandjean, E. Fatigue in industry. *Br. J. Ind. Med.* **1979**, *36*, 75–186. [CrossRef]
9. Lal, S.K.L.; Craig, A. A critical review of the psychophysiology of driver fatigue. *Biol. Psychol.* **2001**, *55*, 173–194. [CrossRef]
10. Jap, B.T.; Lal, S.; Fischer, P.; Bekiaris, E. Using EEG spectral components to assess algorithms for detecting fatigue. *Expert Syst. Appl.* **2009**, *36*, 2352–2359. [CrossRef]
11. Alghowinem, S.; Goecke, R.; Wagner, M.; Parker, G.; Breakspear, M. Eye movement analysis for depression detection. In Proceedings of the 20th IEEE International Conference on Image Processing (ICIP), Melborne, Australia, 15–18 September 2013; pp. 4220–4224.
12. Lee, B.G.; Jung, S.J.; Chung, W.Y. Real-time physicological and vision monitoring of vehicle driver for non-intrusive drowsiness detection. *IET Commun. 2011 5*, 2461–2469.
13. Hess, E.H. Attitude and pupil size. *Sci. Am.* **1965**, *212*, 46–54. [CrossRef] [PubMed]
14. Sampei, K.; Ogawa, M.; Torres, C.C.C.; Sato, M.; Miki, N. Mental Fatigue Monitoring Using a wearable transparent eye detection system. *Micromachines* **2016**, *7*, 20. [CrossRef]
15. Horiuchi, R.; Ogasawara, T.; Miki, N. Fatigue evaluation by detecting blink behavior using eyeglass-shaped optical sensor system. *Micro Nano Lett.* **2017**, *12*, 554–559. [CrossRef]

16. Shigeoka, T.; Ninomiya, T.; Muro, T.; Miki, N. Wearable pupil position detection system utilizing dye-sensitized photovoltaic devices. *Sens. Actuators A Phys.* **2008**, *145–146*, 103–108. [CrossRef]
17. Ozawa, M.; Sampei, K.; Torres, C.C.C.; Ogawa, M.; Oikawa, A.; Miki, N. Wearable line-of-sight detection system using micro-fabricated transparent optical sensors on eyeglasses. *Sens. Actuators A Phys.* **2014**, *205*, 208–214. [CrossRef]
18. Torres, C.C.C.; Sampei, K.; Ogawa, M.; Ozawa, M.; Miki, N. Crosstalk analysis, its effects and reduction techniques among photovoltaic devices used as transparent optical sensors for a wearable line-of-sight detection system. *Jpn. J. Appl. Phys.* **2015**, *54*, 06FP16. [CrossRef]
19. Sugimoto, K.; Kanai, A.; Shoji, N. The effectiveness of the Uchida-Kraepelin test for psychological stress: An analysis of plasma and salivary stress substances. *BioPsychoSoc. Med.* **2009**, *3*, 5. [CrossRef] [PubMed]
20. Kubo, T.; Tachi, N.; Takeyama, H.; Ebara, T.; Inoue, T.; Takanishi, T.; Arakomo, Y.; Murasaki, G.I.; Itani, T. Characteristic patterns of fatigue feelings on four simulated consecutive night shifts by "Jikaku-sho shirabe". *Sangyo Eiseigaku Zasshi* **2008**, *50*, 133–144. (In Japanese) [PubMed]
21. Kubo, T.; Takahashi, M.; Sato, T.; Sasaki, T.; Oka, T.; Iwasaki, K. Weekend sleep intervention for workers with habitually short sleep periods. *Scand. J. Work Environ. Health* **2011**, *37*, 418–426. [CrossRef] [PubMed]
22. Sasaki, T.; Matsumoto, S. Actual conditions of work, fatigue and sleep in non-employed, home-based female information technology workers with preschool children. *Ind. Health* **2005**, *43*, 142–150. [CrossRef] [PubMed]
23. Tanaka, Y. Arousal level and blink activity. *Jpn. J. Psychol.* **1999**, *70*, 1–8. (In Japanese) [CrossRef]

micromachines

MDPI

Article

Adenosine Triphosphate Measurement in Deep Sea Using a Microfluidic Device

Tatsuhiro Fukuba [1,*], Takuroh Noguchi [2], Kei Okamura [2] and Teruo Fujii [3]

[1]   Japan Agency for Marine-Earth Science and Technology, 2-15 Natsushima–cho, Yokosuka, Kanagawa 237-0061, Japan

[2]   Research and Education Faculty, Kochi University, B200 Monobe, Nankoku, Kochi 783-8502, Japan; noguchitk@kochi-u.ac.jp (T.N.); okamurak@kochi-u.ac.jp (K.O.)

[3]   Institute of Industrial Science, The University of Tokyo, 4-6-1 Komaba, Meguro–ku, 153-8505 Tokyo, Japan; tfujii@iis.u-tokyo.ac.jp

*   Correspondence: bafuk@jamstec.go.jp; Tel.: +81-46-867-9374

Received: 10 May 2018; Accepted: 24 July 2018; Published: 27 July 2018

Abstract: Total ATP (adenosine triphosphate) concentration is a useful biochemical parameter for detecting microbial biomass or biogeochemical activity anomalies in the natural environment. In this study, we describe the development and evaluation of a new version of in situ ATP analyzer improved for the continuous and quantitative determination of ATP in submarine environments. We integrated a transparent microfluidic device containing a microchannel for cell lysis and a channel for the bioluminescence L–L (luciferin–luciferase) assay with a miniature pumping unit and a photometry module for the measurement of the bioluminescence intensity. A heater and a temperature sensor were also included in the system to maintain an optimal temperature for the L–L reaction. In this study, the analyzer was evaluated in deep sea environments, reaching a depth of 200 m using a remotely operated underwater vehicle. We show that the ATP analyzer successfully operated in the deep-sea environment and accurately quantified total ATP within the concentration lower than 5×10^{-11} M.

Keywords: ATP; microfluidic device; luciferin–luciferase assay

---

1. Introduction

The importance of organic and inorganic matter circulating globally and locally in ocean environments has prompted research in the field of marine environmental microbiology on the abundance, distribution and roles of oceanic microbes represented by *Eubacteria* and *Archaea*. Microbes have relevant roles especially in submarine hydrothermal sites or hydrocarbon seepage areas, because they support unique ecosystems as primary producers [1]. Even with the rapid progress of sophisticated DNA and RNA analysis methodologies, the determination of the number of microbial cells in seawater samples is still indispensable for estimating microbial biomass and for studying their spatiotemporal distribution. Generally, microscopic or flow-cytometric counting of fluorescently stained or genetically labeled cells are conducted by well-trained researchers in onboard or onshore laboratories, using samples collected during scientific cruises [2]. However, the quantitative determination of microbial ATP (adenosine triphosphate), which is a ubiquitous biomolecule utilized for energy conversion and storage in living cells, in seawater has been regarded as one of the most useful alternatives to labor-intensive microscopic cell enumerations [3,4]. In particular, the quantity of particulate ATP (pATP) in seawater is a representative proxy of the microbial biomass in a sample [5]. Since dissolved ATP (dATP)—an important carbon and phosphorus source for marine microbes—is also related to microbial activity [6], the sum of pATP and dATP (total ATP (tATP)) is a useful parameter indicative of the presence of biogeochemical events, such as submarine volcanisms [1], hydrocarbon seepages [7] and occasional supply of organic resources (e.g., whale falls) [8]. By realizing a compact and portable apparatus for in situ ATP quantification, it becomes possible to analyze the distribution of microbial biomass anomalies with an unprecedented spatiotemporal resolution, which enables an efficient exploration of the underwater resources as well as deep sea environmental and microbiological studies [9].

ATP concentration can be determined by the L–L (luciferin–luciferase) bioluminescence assay [10] (see Figure 1) that is a simple method for ATP quantification generally used for hygiene monitoring [11,12]. The microfluidic technology has been applied to automate flow analyses in various fields, including biochemistry and it has been used in marine environments for microbial gene analysis [13], nutrient analysis [14] and trace metal analysis [15,16]. Previously, we developed and evaluated microfluidic devices and in situ analyzers for intermittent (non-continuous) quantitative determination of ATP based on the L–L assay [17–19]. For the further improvement of the spatiotemporal resolution of the in-situ measurement, a new system with continuous measurement capability was developed and evaluated for practical oceanography applications [20–22]. In these studies, two microfluidic devices with single function for microbial cell lysis and bioluminescence detection were used in combination. In this study, we improve the new system by merging the single function microfluidic devices into one device and its performance in the deep-sea environments is examined by comparing the in-situ ATP quantification results with manually processed values.

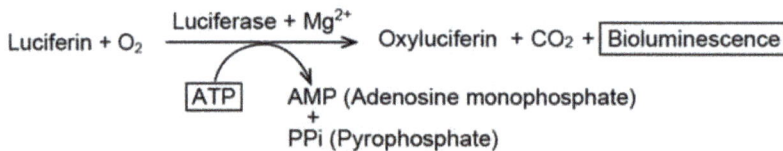

$$\text{Luciferin} + O_2 \xrightarrow{\text{Luciferase} + Mg^{2+}} \text{Oxyluciferin} + CO_2 + \boxed{\text{Bioluminescence}}$$

$$\boxed{\text{ATP}} \quad \text{AMP (Adenosine monophosphate)} \\ + \\ \text{PPi (Pyrophosphate)}$$

Figure 1. Schematic of luciferin–luciferase reaction for ATP quantification.

2. Materials and Methods

2.1. In situ ATP Analyzer

The in-situ ATP analyzer that we have produced can quantify the ATP contained in seawater continuously at a depth of 3000 m. Since the seawater samples are not filtered before the analysis, tATP concentrations are measured. The analyzer consists of an analysis module, which is the core component and of a photometry module for the bioluminescence intensity measurements based on the L–L reaction (see Figure 2).

A flow diagram of the ATP analyzer is shown in Figure 3. The analysis module (see Figure 4) is connected to the microfluidic device for the L–L assay (Figures 2 and 5) and contains three miniature peristaltic pumps (RP-0.15S-P15A-DC5VS, Aquatech Co., Ltd., Daito, Japan), three solenoid-actuated three-way valves (STV-3-1, Takasago Electric, Inc., Nagoya, Japan) and the control electronics. The fluidic components are connected using black-colored Teflon™ FEP (fluorinated ethylene propylene) tubes (1519, IDEX Health & Science, Oak Harbor, WA, USA) to shield the analysis module from the ambient light. A cascaded connection of the three-way valves enables the selection of four kinds of fluids (the sample and three standard solutions) through the three valves. The control electronics is based on a miniature microprocessor board (ML100 series, Microtec Co., Ltd., Funabashi, Japan) and is used to control the valves, pumps and heater. Sequential scenarios for pump and valve operation and temperature setting can be stored on a micro SD card in the control electronics. All the components of the analysis module are enclosed in a cylindrical plastic container filled with fluorinated oil (Fluorinert FC-43, 3M, Maplewood, MN, USA) for electrical insulation and pressure equalization during underwater operations. Since it employs a pressure-balanced configuration, the system does not require a complicated and large pumping mechanism to manage the elevated underwater hydrostatic pressure and overcome the differences between the internal and external pressures. The seawater sample is led from the outside of the system through a short tube to the container. The chemicals for the L–L assay and standard solutions are stored in plastic bags, connected to the system via the Teflon™ FEP tube. One end of the oil-filled analysis module has a transparent window facing the photometry module. The total power consumption of the in-situ ATP analyzer is approximately 24 W in maximum (when all the valves and the heater are activated) including the photometry module described later.

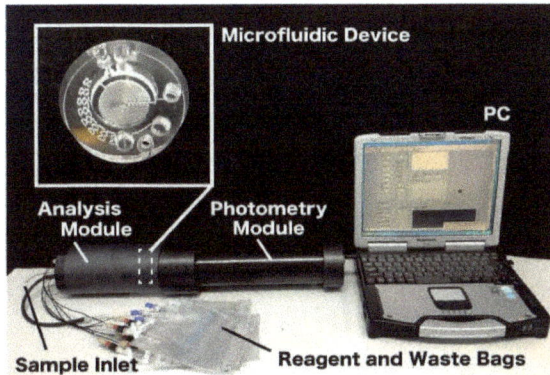

Figure 2. The in-situ adenosine triphosphate (ATP) analyzer and the microfluidic device developed and evaluated in this study with a laptop PC for real-time control and data acquisition.

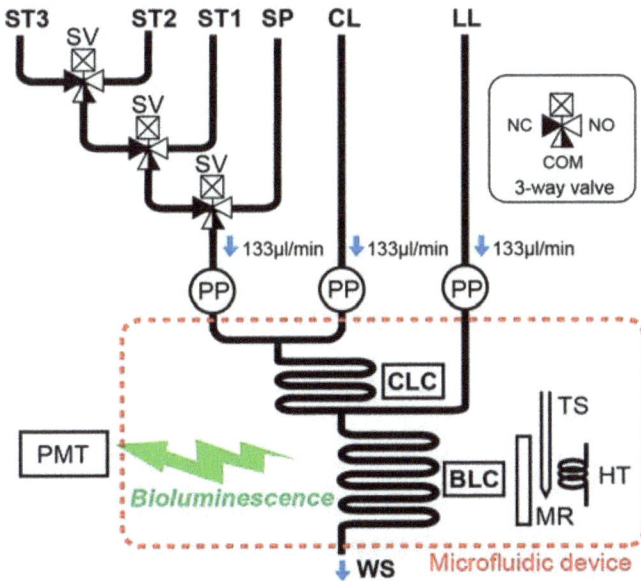

Figure 3. Flow diagram of the in-situ ATP analyzer. ST1–3: ATP standard solutions 1–3, SP: sample, CL: cell lysis reagent, LL: L–L reagent, CLC: cell lysis channel, BLC: bioluminescence channel, WS: waste outlet, PP: peristaltic pump, SV: solenoid three-way valve, TS: temperature sensor, HT: heater, MR: mirror, PMT: photomultiplier tube.

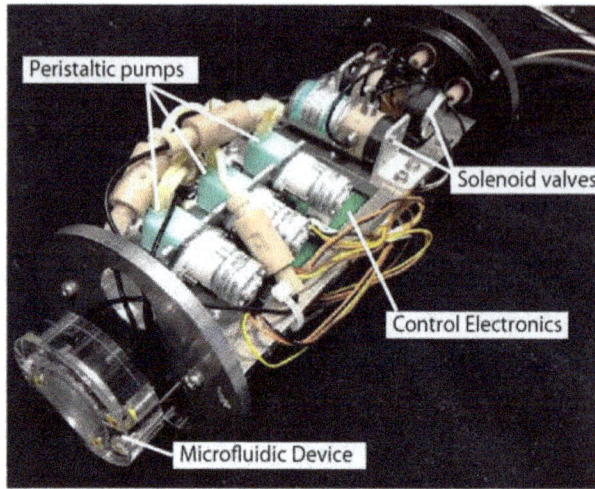

Figure 4. The analysis module for the in-situ ATP analyzer. Three peristaltic pumps and three solenoid-actuated three-way valves were employed as the pumping device.

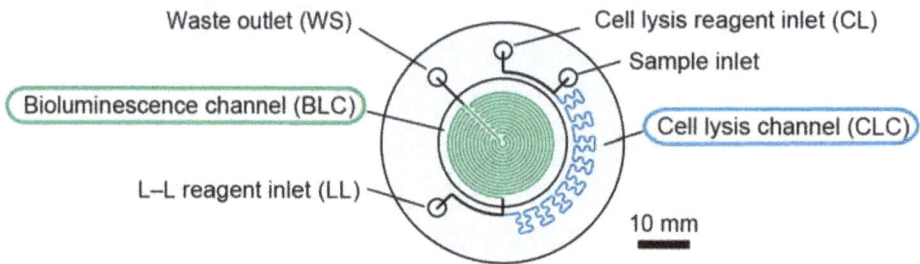

Figure 5. Microfluidic device for the in-situ ATP analyzer.

2.2. Microfluidic Device

The microfluidic device for the in-situ ATP analyzer is made of transparent PMMA (Poly (methyl methacrylate)) to detect the bioluminescence produced by the L–L reaction taking place in the device (Figure 5). Channels for cell lysis and the L–L reaction are engraved by means of precision milling on a PMMA disk (47 mm in diameter and 2 mm in thickness) and bonded with the top plate (11 mm in thickness) with 1/4-28 UNF threaded tubing interfaces by a diffusion bonding method (Takasago Electric., Inc., Nagoya, Japan). Each channel is 0.5 mm wide and 1.5 mm deep. First, the sample (SP) is mixed with a cell lysis (CL) reagent that releases ATP from the microbial cells as the mixture passes through a serpentine cell lysis channel (CLC), which is approximately 133 mm in length (this takes approximately 23 s). The extracted ATP is mixed with the L–L reagent immediately at the end of the CLC to initiate the L–L bioluminescence reaction that takes place in the serpentine bioluminescence channel (BLC) folded in a circular shape. Along the approximately 482 mm of the BLC, bioluminescence is emitted and its intensity corresponds to the ATP concentration in the sample. Even though the microfluidic device does not include a mixer structure, the asynchronous pulsated flow generated by the three peristaltic pumps enables the mixing of the reagents. The mixing ratio of the sample and reagents is 1:1:1 (SP/CL/LL). At the flow rate of 133 µL/min for each component, it takes approximately 54 s for the final mixture to pass through the L–L reaction channel. The waste is discharged from a waste port (WS) and collected in a plastic bag outside of the analysis module. A miniature flat mirror (MR) reflects the bioluminescence to a PMT (photomultiplier tube) in the photometry module. A film heater (HT) and a temperature sensor (TS) are fixed behind the mirror in order to maintain the optimal temperature for the L–L reaction. The TS was placed just behind the HT and the activation of the HT was regulated by the control electronics in the analysis module. The temperature is typically kept to 35 °C (higher than room-temperature) considering the situation of operation under the room-temperature condition with the waste heat generation from pumps and valves in the analysis module.

2.3. Photometry Module

The photometry module consists of a photon-counting PMT and a data logging electronics located in a cylindrical pressure-tight housing. In contrast to the analysis module that has a pressure-balanced configuration, all the key components of the photometry module are protected from ambient hydrostatic pressure. The PMT faces a pressure-resistant glass window (18 mm in thickness) fixed at the end of the pressure-tight housing for the bioluminescence measurement. The measured bioluminescence intensity data are stored on a micro SD card in the data logging electronics based on ML100 series (Microtec Co., Ltd., Funabashi, Japan) and transferred in real-time to a PC connected to the in-situ ATP analyzer via RS-232 format.

2.4. Reagents

For the L–L assay, a commercially available kit for bacterial biomass determination (CheckLite HS Set, Kikkoman Biochemifa Co., Tokyo, Japan) containing the ATP releasing (cell lysis) reagent and the L–L reagent was modified for seawater sample measurement [18,19]. EDTA (ethylenediaminetetraacetic acid, Wako Pure Chemical Industries, Ltd., Osaka, Japan) was added to the ATP releasing reagent at a final concentration of 10 mM to avoid precipitation in the presence of the seawater samples or the seawater-based ATP standard solutions. To avoid the adsorption of the reagents and the adhesion of natural particles or debris to the micro-channels, 2% (v/v) Tween 20 (MP Biomedicals LLC, Santa Ana, CA, USA) was added to the L–L reagent. Both EDTA and Tween 20 were sterilized by autoclaving prior to use to eliminate potentially contaminating ATP. Seawater-based ATP standard solutions (5×10^{-12}, 5×10^{-11}, 5×10^{-10} M ATP and blank) were prepared from an original ATP standard solution (2×10^{-6} M ATP, Kikkoman Biochemifa Co., Tokyo, Japan) by diluting it with autoclaved artificial seawater (Daigo's artificial seawater SP for marine microalgae medium, Nihon Pharmaceutical Co., Ltd., Tokyo, Japan). All reagents and standards were aseptically introduced into sterilized plastic bags (DSF-300, Tsukada Medical Research Co. Ltd., Ueda, Japan) for use in the in-situ analyzer.

2.5. Evaluations

The in-situ ATP analyzer was evaluated in the laboratory environment using three ATP standard solutions and blank to acquire the relationship between ATP concentration and bioluminescence intensity. For the evaluation, ATP standard solutions and blank were filled in an aseptic plastic test tube and introduced into the analyzer from the sample inlet. After the saturation of the bioluminescence intensity at each ATP concentration, consecutive 10 s values were used for calibration.

2.6. In Situ Evaluations

The evaluation of the in-situ ATP analyzer developed in this study was carried out in the real field during the scientific cruise KS-17-J07C using R/V SHINSEI MARU and ROV (remotely operated vehicle) HYPER-DOLPHIN (Japan Agency for Marine-Earth Science and Technology, JAMSTEC) in May 2017. The in-situ ATP analyzer was mounted on the ROV (see Figure 6), which provided electricity and RS-232 communication. Real-time monitoring of the bioluminescence data and control of the in-situ ATP analyzer were carried out on board using a PC, which was connected to the ROV via an underwater cable. During the 2021st dive of HYPER-DOLPHIN, the ROV and the analyzer were dived to the Oomuro Hole located in the northern Izu-Ogasawara arc, where the existence of hydrothermal activity has been reported [23]. Continuous tATP measurements were performed from the sea surface to the seafloor at a depth of approximately 200 m. The system calibration was also carried out in situ using the ATP standard solutions to compensate the changes on the reagent flow-rate or temperature on the microfluidic device caused by unexpected effect of elevated hydrostatic pressure on the pumps, temperature sensor and heater controller. Furthermore, effect of the temperature and pressure on the L–L reaction [24] must be considered for accurate in situ quantification of tATP. Water samplers (see Figure 6c) equipped with aseptic plastic syringes were installed on the ROV for sample collection for data comparison. The collected water samples were transferred to clean test tubes immediately after the dive and tATP concentration was measured by a conventional method using test tubes on board using a desktop luminometer (NU-2600, Microtec Co., Ltd., Funabashi, Japan) and a CheckLite HS Set (Kikkoman Biochemifa Co., Tokyo, Japan) without modification. Bioluminescence intensity was integrated for 10 s and all the measurements were triplicated.

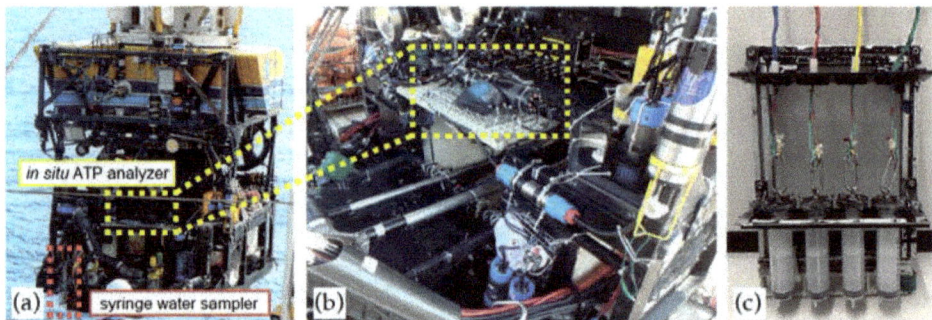

Figure 6. Remotely operated vehicle (ROV) HYPER-DOLPHIN with the in-situ ATP analyzer (**a**); Close-up view of the in-situ ATP analyzer mounted on the payload space of the ROV (**b**); The syringe water sampler (**c**).

3. Results

As a result of the calibration in the laboratory environment, highly linear relationship ($R^2 > 0.99$) between the ATP concentration and the bioluminescence intensity was obtained (Figure 7). Therefore, extrapolation of the result of in situ calibrations obtained using the $5 \times 10^{-12}, 5 \times 10^{-11}$ M of ATP standards and the blank is reasonable up to 5×10^{-10} M of ATP concentration.

A continuous tATP measurement from the sea surface to the bottom of the Oomuro Hole area was carried out successfully using the in-situ ATP analyzer developed in this study. After calibrating the system using ATP standard solutions, a linear correlation between the ATP concentration and the bioluminescence intensity ($R^2 > 0.99$) was obtained (see Figure 8) and was later applied to the measured raw data to calculate tATP concentration in the samples (see Figure 9). For data exceeding the bioluminescence intensity corresponding to the ATP concentration of 5.0×10^{-11} M, such as data from the surface, the calibration formula was extrapolated.

At the beginning of the measurement at the surface, the measured tATP concentration was extraordinarily low and increased rapidly within five minutes. This was due to a time lag at the beginning of the measurement required to reach and fill the reagents and the sample to the microfluidic device. A high concentration of tATP, corresponding to 1.0×10^{-10} M or more, was measured at the surface after the reagents and sample were filled in the microfluidic device. This is consistent with the formation of a larger microbial biomass layer at the surface supported by photosynthetic primary productions. As the ROV dived more deeply, the tATP concentration decreased rapidly. In situ calibration was successfully performed at the bottom of the sea, as shown in Figure 9. The measurement precision rates of the analyzer, estimated from the 3σ value calculated from three consecutive 10 s measurements of bioluminescence intensity of the 5.0×10^{-12} and 5.0×10^{-11} M ATP standard solutions, were 43% and 4.5% (2.2×10^{-12} and 2.3×10^{-12} M) of the measured values, respectively. Conversely, the measurement precision rate calculated for manually measured triplicate data were 14% and 9.9% of the measured values. The in-situ ATP analyzer showed better performance for the determination of ATP concentration close to 5×10^{-11} M. In contrast, the measurement precision rate for lower ATP concentrations was better for the desktop apparatus. This was due to fluctuations of the bioluminescence intensity data during the measurement of the 5×10^{-12} M standard.

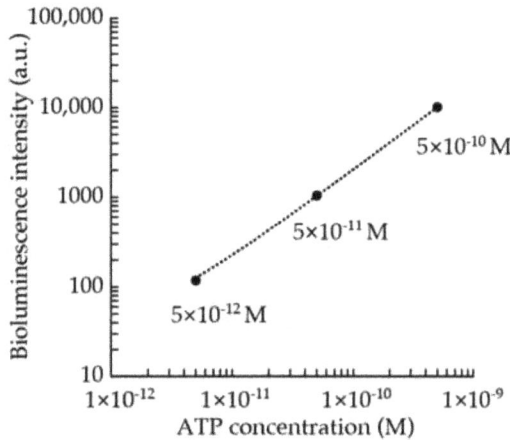

Figure 7. Relationship between ATP concentration and bioluminescence intensity measured by the in-situ ATP analyzer in the laboratory environment. Continuous data for 10 s at each ATP concentration were integrated, the blank value was subtracted and plotted with the ATP concentration.

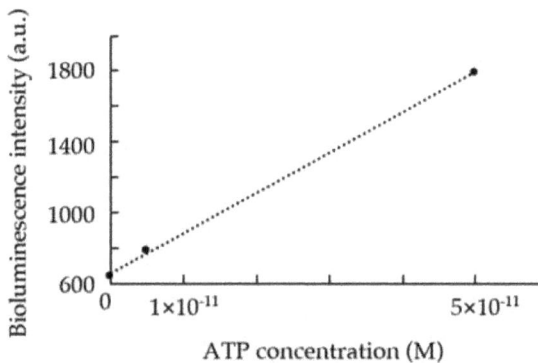

Figure 8. Relationship between ATP concentration and bioluminescence intensity measured by the in-situ ATP analyzer during the 2021st dive of HYPER-DOLPHIN. Consecutive 10 s measurements at each ATP concentration were integrated and plotted with the ATP concentration.

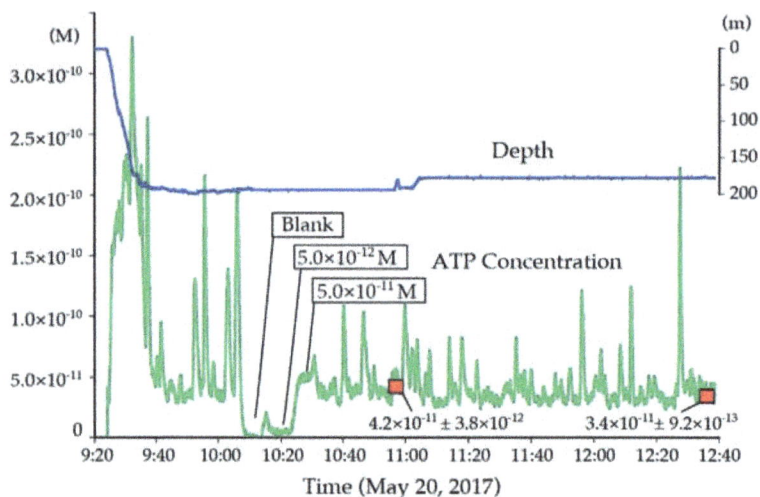

Figure 9. Result of in situ ATP measurement at the Oomuro Hole during the 2021st dive of HYPER-DOLPHIN. Raw bioluminescence data were converted to ATP concentration and plotted as the green line with time. The time was shifted 3 min ahead considering the time-lag between sample intake and bioluminescence emission in the analyzer. The depth profile measured by a depth sensor on the ROV is shown in the blue line. ATP concentrations measured onboard using the collected water samples were shown as red squares.

In the Oomuro Hole, tATP concentration was in the range of 2.0 to 3.0×10^{-11} M with occasional increases to values higher than 1.0×10^{-10} M. The ATP concentrations measured on board of two seawater samples collected by the ROV were consistent with the data provided by the in-situ ATP analyzer, as shown in Figure 9. The occasional ATP concentration peaks were likely due to the introduction of inorganic-organic aggregated particles including microbes [25] or marine snow particles originated from the surface water. After more than 2 h of the operation of the in-situ ATP analyzer was halted because of an electric trouble alert on the ROV system.

4. Discussion

In this study, an in-situ ATP analyzer was developed by employing a PMMA microfluidic device as a core element of the system. The performance of the in-situ ATP analyzer was evaluated in a real deep-sea environment. The ATP analyzer successfully measured the tATP concentrations at different depths, providing data that were consistent with those measured manually. These results demonstrate that a portable, simple and reliable flow analysis system such as our microfluidic device can be used in extreme environments for real-time biochemical analyses. The calculated measurement precision rates showed successful value (4.5%) at 5×10^{-11} M range of ATP concentration and decreased performance for the determination of ATP concentrations as low as 5×10^{-12} M (43%). Because the lower measurement precision has been led by fluctuation of light intensity value and it is caused by electric noise from control electronics, improvement in the precision may be achieved by the reducing the noise by enhancing the control electronics in the near future.

The pATP concentration required for the quantitative estimation of the microbial biomass can be determined using the current system by subtracting dATP concentration, measured by the L–L assay without using the cell lysis reagent, from the tATP concentration [26]. However, to measure the dATP, it is necessary to perform additional calibrations of the in-situ ATP analyzer, specific for dATP measurements. In the current system, the ATP measurement must be interrupted during the system calibration with ATP standard solutions. The calibration process required approximately 30 min during the in-situ evaluation at the Oomuro Hole. To overcome this limitation, we have been developing and evaluating a novel calibration method using an optically activated caged ATP as an internal standard [21,22,26]. By applying the new calibration technology, it will be possible to utilize the in-situ ATP analyzer for practical underwater resources surveys and environmental assessment missions in the near future.

Instruments for in situ flow-analysis based on the technologies developed in this study and consisting of a simple microfluidic device and a pumping apparatus can be employed for various in situ biological and biochemical analyses in human-inaccessible extreme environments.

Author Contributions: T.F. (Tatsuhiro Fukuba) took part in the conceptualization of the study. T.F. (Tatsuhiro Fukuba) took part in device design and instrumentation. K.O. and T.N. took part in cruise planning and management. T.F. (Tatsuhiro Fukuba) wrote and prepared the original draft. T.F. (Tatsuhiro Fukuba) and T.N. wrote, reviewed and edited the manuscript. T.F. (Teruo Fujii) supervised the study. T.F. (Teruo Fujii) and K.O. acquired funding.

Funding: This research was funded by the Council for Science, Technology and Innovation (CSTI), Cross-ministerial Strategic Innovation Promotion Program (SIP) "Next-generation technology for ocean resources exploration" (Lead agency: JAMSTEC) and Ministry of Education, Culture, Sports, Science and Technology (MEXT), "Development of new tools for the exploration of seafloor resources" project.

Acknowledgments: The authors are grateful to the crew of the R/V SHINSEI-MARU and the operating team of HYPER-DOLPHIN (JAMSTEC) for their helpful assistance during the scientific cruise KS-17-J07C.

Conflicts of Interest: The authors declare no conflict of interest.

References

1. Karl, D.M.; Wirsen, C.O.; Jannasch, H.W. Deep-Sea primary production at the Galápagos hydrothermal vents. *Science* **1980**, *207*, 1345–1347. [CrossRef]

2. Monger, B.C.; Landry, M.R. Flow cytometric analysis of marine bacteria with Hoechst 33342. *Appl. Environ. Microbiol.* **1993**, *59*, 905–911. [PubMed]

3. Holm-Hansen, O.; Booth, C.R. The measurement of adenosine triphosphate in the ocean and its ecological significance. *Limnol. Oceanogr.* **1966**, *11*, 510–519. [CrossRef]

4. Karl, D.M. Cellular nucleotide measurements and applications in microbial ecology. *Microbiol. Rev.* **1980**, *44*, 739–796. [PubMed]

5. Daffonchio, D.; Borin, S.; Brusa, T.; Brusetti, L.; van der Wielen, P.W.J.J.; Bolhuis, H.; Yakimov, M.M.; D'Auria, G.; Giuliano, L.; Marty, D.; et al. Stratified prokaryote network in the oxic-anoxic transition of a deep-sea halocline. *Nature* **2006**, *440*, 203–207. [CrossRef] [PubMed]

6. Björkman, K.M.; Karl, D.M. A novel method for the measurement of dissolved adenosine and guanosine triphosphate in aquatic habitats: Applications to marine microbial ecology. *J Microbiol. Meth.* **2001**, *47*, 159–167. [CrossRef]

7. Sibuet, M.; Olu, K. Biogeography, biodiversity and fluid dependence of deep-sea cold-seep communities at active and passive margins. *Deep Sea Res. Part II* **1998**, *45*, 517–567. [CrossRef]

8. Treude, T.; Smith, C.R.; Wenzhöfer, F.; Carney, E.; Bernardino, A.F.; Hannides, A.K.; Krüger, M.; Boetius, A. Biogeochemistry of a Deep-sea whale fall: Sulfate reduction, sulfide efflux and methanogenesis. *Mar. Ecol. Prog. Ser.* **2009**, *382*, 1–21. [CrossRef]

9. Fukuba, T.; Miwa, T. Novel sensors and platforms for monitoring of deep-sea environment. *Aquabiology* **2016**, *38*, 131–137. (in Japanese with English abstract).

10. Ludin, A. Use of firefly luciferase in ATP-related assays of biomass, enzymes and metabolites. *Meth. Enzymol.* **2000**, *305*, 346–370. [CrossRef]

11. Bautista, D.A.; McIntyre, L.; Laleye, L.; Griffiths, M.W. The application of ATP bioluminescence for the assessment of milk quality and factory hygiene. *J. Rapid Methods Autom. Microbiol.* **1992**, *1*, 179–193. [CrossRef]

12. Hawronskyj, J.M.; Holah, J. ATP: A universal hygiene monitor. *Trends Food Sci. Tech.* **1997**, *8*, 79–84. [CrossRef]

13. Fukuba, T.; Miyaji, A.; Okamoto, T.; Yamamoto, T.; Kaneda, S.; Fujii, T. Integrated in situ genetic analyzer for microbiology in extreme environments. *RSC Adv.* **2011**, *1*, 1567–1573. [CrossRef]

14. Beaton, A.D.; Cardwell, C.L.; Thomas, R.S.; Sieben, V.J.; Legiret, F.-E.; Waugh, E.M.; Statham, P.J.; Mowlwm, M.C.; Morgan, H. Lab-on-a-chip measurement of nitrate and nitrite for *in situ* analysis of natural waters. *Environ. Sci. Tech. Lett.* **2012**, *46*, 9548–9556. [CrossRef] [PubMed]

15. Chapin, T.P.; Jannasch, H.W.; Johnson, K.S. In situ osmotic analyzer for the year-long continuous determination of Fe in hydrothermal systems. *Anal. Chim. Acta* **2002**, *463*, 265–274. [CrossRef]

16. Provin, C.; Fukuba, T.; Okamura, K.; Fujii, T. An integrated microfluidic system for manganese anomaly detection based on chemiluminescence: Description and practical use to discover hydrothermal plumes near the Okinawa Trough. *IEEE J. Ocean Eng.* **2012**, *38*, 178–185. [CrossRef]

17. Aoki, Y.; Fukuba, T.; Yamamoto, T.; Fujii, T. Design optimization and evaluation of a bioluminescence detection part on a microfluidic device for in situ ATP quantification. *IEEJ Trans. Sens. Micromach.* **2009**, *129*, 73–76. [CrossRef]

18. Fukuba, T.; Fujii, T. Bioluminescence detection for ATP quantification using microfluidic device. In *Molecular Biological Technologies for Ocean Sensing*; Tiquia-Arashiro, S., Ed.; Humana Press: Totowa, NJ, USA, 2012; pp. 203–217. ISBN 978-1-61779-914-3.

19. Fukuba, T.; Aoki, Y.; Fukuzawa, N.; Yamamoto, T.; Kyo, M.; Fujii, T. A microfluidic *in situ* analyzer for ATP quantification in ocean environments. *Lab Chip* **2011**, *11*, 3508–3515. [CrossRef] [PubMed]

20. Fukuba, T.; Noguchi, T.; Okamura, K.; Kyo, M.; Nishida, S.; Miwa, T.; Fujii, T. ATP sensing in deep-sea environments using continuous flow microfluidic device. In Proceedings of the 18th International Conference on Miniaturized Systems for Chemistry and Life Sciences (μTAS 2014), San Antonio, TX, USA, 26–30 October 2014; pp. 1912–1914.

21. Hanatani, K.; Fukuba, T.; Fujii, T. Development of in situ microbial ATP analyzer and internal standard calibration method. In Proceedings of the 2015 IEEE Underwater Technology (UT), Chennai, India, 23–25 February 2015. [CrossRef]

22. Hanatani, K.; Fukuba, T.; Okamura, K.; Fujii, T. In situ calibration system for deep-sea microbial ATP sensor using caged ATP. In Proceedings of the 2016 JSPE Spring Conference, Chiba, Japan, 15 March 2016. (in Japanese with English abstract). [CrossRef]

23. Tani, K.; Ishizuka, O.; Nichols, A.R.; Hirahara, Y.; Carey, R.; McIntosh, I.M.; Masaki, Y.; Kondo, R.; Miyairi, Y. Discovery of an active shallow submarine silicic volcano in the northern Izu-Bonin arc: Volcanic structure and potential hazards of Oomurodashi volcano. In Proceedings of the American Geophysical Union, Fall Meeting 2013, San Francisco, CA, USA, 9–13 December 2013.

24. Ueda, I.; Shinoda, F.; Kamaya, H. Temperature-dependent effects of high pressure on the bioluminescence of firefly luciferase. *Biophys. J.* **1994**, *66*, 2107–2110. [CrossRef]

25. Toner, B.M.; Fakra, S.C.; Manganini, S.J.; Santelli, C.M.; Marcus, M.A.; Moffett, J.W.; Rouxel, O.; German, C.R.; Edwards, K.J. Presence of iron (II) by carbon-rich matrices in a hydrothermal plume. *Nat. Geosci.* **2009**, *2*, 197–201. [CrossRef]

26. Fukuba, T.; Hanatani, K.; Okamura, K.; Fujii, T. Microfluidic device for in situ quantification of marine microbial ATP with in-line photolysis of caged ATP as internal standard. In Proceedings of the 20th International Conference on Miniaturized Systems for Chemistry and Life Sciences (μTAS 2016), Dublin, Ireland, 9–13 October 2016; pp. 1469–1470.

MDPI

St. Alban-Anlage 66

4052 Basel

Switzerland

Tel. +41 61 683 77 34

Fax +41 61 302 89 18

www.mdpi.com

Micromachines Editorial Office

E-mail: micromachines@mdpi.com

www.mdpi.com/journal/micromachines

www.ingramcontent.com/pod-product-compliance
Lightning Source LLC
Chambersburg PA
CBHW051906210326
41597CB00033B/6046